Lecture Notes in Computer Scie

Commenced Publication in 1973
Founding and Former Series Editors:
Gerhard Goos, Juris Hartmanis, and Jan van Leeuwen

Christoph Bussler   Dieter Fensel
Maria E. Orlowska   Jian Yang (Eds.)

Web Services,
E-Business,
and the Semantic Web

Second International Workshop, WES 2003
Klagenfurt, Austria, June 16-17, 2003
Revised Selected Papers

 Springer

Volume Editors

Christoph Bussler
National University of Ireland, Digital Enterprise Research Institute (DERI)
University Road, Galway, Ireland
E-mail: chris.bussler@deri.ie

Dieter Fensel
University of Innsbruck, Digital Enterprise Research Institute (DERI)
Technikerstr. 13, 6020 Innsbruck, Austria
E-mail: dieter.fensel@deri.org

Maria E. Orlowska
The University of Queensland
School of Information Technology and Electrical Engineering
St. Lucia, Brisbane, QLD 4072, Australia
E-mail: maria@itee.uq.edu.au

Jian Yang
Tilburg University, Infolab
P.O. Box 90153, 5000 LE Tilburg, The Netherlands
E-mail: jian@uvt.nl

Library of Congress Control Number: 2004109713

CR Subject Classification (1998): H.3, H.4, H.5.3, H.2, K.4.3-4

ISSN 0302-9743
ISBN 3-540-22396-7 Springer Berlin Heidelberg New York

Springer is a part of Springer Science+Business Media

springeronline.com

© Springer-Verlag Berlin Heidelberg 2004
Printed in Germany

Typesetting: Camera-ready by author, data conversion by Scientific Publishing Services, Chennai, India
Printed on acid-free paper   SPIN: 11019770   06/3142   5 4 3 2 1 0

Preface

The 2nd Workshop on Web Services, E-Business, and the Semantic Web (WES) was held during June 16–17, 2003 in conjunction with CAiSE 2003, the 15th International Conference on Advanced Information Systems Engineering.

The Internet is changing the way businesses operate. Organizations are using the Web to deliver their goods and services, to find trading partners, and to link their existing (maybe legacy) applications to other applications. Web services are rapidly becoming the enabling technology of today's e-business and e-commerce systems, and will soon transform the Web as it is now into a distributed computation and application framework.

On the other hand, e-business as an emerging concept is also impacting software applications, the everyday services landscape, and the way we do things in almost each domain of our life. There is already a body of experience accumulated to demonstrate the difference between just having an online presence and using the Web as a strategic and functional medium in e-business-to-business interaction (B2B) as well as marketplaces.

Finally, the emerging Semantic Web paradigm promises to annotate Web artifacts to enable automated reasoning about them. When applied to e-services, the paradigm hopes to provide substantial automation for activities such as discovery, invocation, assembly, and monitoring of e-services.

But much work remains to be done before realizing this vision.

Clearly Web services must satisfy a number of challenging requirements in order to be able to play a crucial role in the new application domain of e-business and distributed application development. They should be modeled and designed to reflect the business objectives. Although some progress has been made in the area of Web service description and discovery, and there are some important standards like SOAP, WSDL, and UDDI emerging, there is still a long way to go. There is still a list of issues that need to be addressed and researched in connection with foundations, technology support, modeling methodologies, and engineering principles before Web services becomes the prominent paradigm for distributed computing and electronic business.

The goal of this workshop is to bring Web services, e-business, and Semantic Web technological issues together for discussion and review. This includes new research results and developments in the context of Web services and e-business as well as application of existing research results in this new fascinating area.

Besides very interesting and stimulating research paper presentations, two keynotes were delivered in the morning of every workshop day. Pat Croke addressed the workshop with a keynote titled "Enterprise Application Integration in 2010 A.D.". He gave interesting insights into past, current and future developments in the space of Enterprise Application Integration.

Robert Meersman reviewed the state of the art with an interesting keynote titled "Old Ontology Wine in New Semantic Bags, and Other Scalability Issues". The keynote gave important insights and created a lively discussion.

We would like to thank the WES program committee for their hard work in helping make this workshop a success.

June 2003

Christoph Bussler
Dieter Fensel
Maria E. Orlowska
Jian Yang

Workshop Organizing Committee

Christoph Bussler
Digital Enterprise Research Institute (DERI), Ireland

Dieter Fensel
Digital Enterprise Research Institute (DERI), Austria and Ireland

Maria E. Orlowska
University of Queensland, ITEE, Australia

Jian Yang
Tilburg University, The Netherlands

Program Committee

Witold Abramowicz
Marco Aiello
Vincenzo D'Andrea
Fabio Casati
Andrzej Cichocki
Dickson Chen
Sing-Chi Cheung
Pat Croke
Umeshwar Dayal
Paul Grefen
Manfred Hauswirth
Patrick Hung
Matthias Jarke
Kamal Karlapalem
Larry Kerschberg
Heiko Ludwig
Massimo Mecella

Borys Omelayenko
Tamer Ozsu
George Papadopoulos
Barbara Pernici
Charles Petrie
Manfred Reichert
Michael Rosemann
Shazia Sadiq
Wasim Sadiq
Karsten Schulz
Ming-Chien Shan
Maarten Steen
Goh Eck Soong
Roger Tagg
Willem-Jan van Heuvel
Gerhard Weikum

Table of Contents

Enterprise Business Integration in 2010A.D.[1]

Pat Croke

Hewlett-Packard, European Software Centre, Galway
Pat.Croke@hp.com

Abstract. This paper looks at how the use of ontologies to describe businesses and systems may allow a move away from standardization, as a basis for Enterprise Business Integration in favor of mediation between different standards. It proposes the Supply Chain Councils SCOR [1] model as a possible base for an ontology to describe businesses and systems. A workbench is described that would allow automatic configuration of orchestration systems to manage business to business and system to system integration. It also shows how the Semantic Web [2] could be used by agents to identify beneficial changes in a company's supply chain.

To quote George Bernard Shaw:

"The reasonable man adapts himself to the world; the unreasonable one persists in trying to adapt the world to himself. Therefore all progress depends on the unreasonable man."

To date Enterprise Business Integration has tended to take the *"reasonable man"* approach attempting to standardize interfaces between systems. It is the premise of this paper that Semantic Web Services technology and in particular ontology based mediation will allow the *"unreasonable man"* approach where systems are free to use different standards.

Enterprise Business Integration or (EBI) refers to both Enterprise Application Integration (EAI) and Business-to-Business integration. It is the largest area of I.T. expenditure on which $3.9 billion will be spent in 2003 increasing to $5.6 in 2006 according to the Aberdeen Group [3]. The need to do EBI is as inevitable as death and taxes because of:

1. The replacement of old systems for purposes of increasing functionality or maintainability.
2. Companies continuously restructuring to optimize their performance through centralizing or decentralizing activities such as purchasing.
3. Companies merging with other companies.
4. Companies divesting themselves of businesses they no longer consider core or profitable. Conversely they acquire new companies where they see potential.
5. Companies changing which activities they wish to do internally and which they want done by a supplier. In-sourcing occurs where an activity previously done

---

[1] Opinions and intuitions expressed in this invited keynote address at CaiSE's workshop on Web Services, E-Business and the Semantic Web, are the author's and do not necessarily reflect Hewlett-Packard Company's position.

C. Bussler et al. (Eds.): WES 2003, LNCS 3095, pp. 1–10, 2004.

by a supplier is brought in-house. Out-sourcing occurs where an activity previously done in-house is carried out by a supplier.
6. Companies changing their distribution strategy. Prior to the Internet a lot of companies used a network of resellers and distributors between them and their end customer. Now they are selling direct to the end customer over the Internet. When a new layer is added this is referred to as intermediation. When a layer is removed it is called disintermediation.

The only companies without a future need to do EBI, are those that have gone out of business. Drivers of cost are: analyzing the impact of a change, and coding/testing the solution.

1 The Unreasonable World

In order to make progress, new methods and technologies are introduced which are better than those that went before. The latest of these is Web Services [4]. New systems under development will take advantage of this technology. However it would be impossible to move all existing systems to use the new technology due to cost. So from an EBI point of view it is just one more technology to be taken care of. This means it will never be possible to use one standard. EBI exists in an unreasonable world.

Businesses use documents such as Purchase Orders or Work Orders to communicate between internal processes, and externally with customers or suppliers. These documents not only convey instructions, but are also the basis of accounting entries within an enterprise's ledgers and sub-ledgers. They are required by law as evidence that a transaction has taken place and are usually required to be retained for a number of years.

Prior to electronic commerce all of these documents were paper based and were as individual as possible to reflect a company's branding and image. This was no problem for humans who are smart and able to select the pieces of information they require from the document.

E-Commerce requires that the electronic formats of these documents be standardized because computers are stupid and the tiniest change in format requires a change to a computer program. An important point is that standardization is only required because of the current low capabilities of computers.

1.1 Formats and Interface Technologies

The number of formats and technologies for communicating between systems continues to increase. These do not tend to replace existing formats. On the contrary they are only used in new system development. They then become yet another technology for the enterprise business integrator to deal with.

There are an ever increasing number of interface technologies used for communicating between systems such as: flat file exchange, messaging, RPC, Sockets, HTTP [5], CORBA [6], DCOM [7], FTP, .NET [8]. There are numerous formats for messages such as: fixed mapped record formats, variable mapped formats, Type Length Value, ASN.1 [9], delimited, XML [10] and so on.

In the area of document formats the world continues to become more unreasonable. With EDI there were two main standards for documents ANSI X.12 [11] and EDIFACT [12]. There are an exploding number of XML document formats:. RosettaNet [13], UBL [14], CommerceOne [15], Ariba [16], OAGIS [17], to name some of the more prominent.

1.2 Optional Documents

Even when businesses agree on the format of the messages they will use to exchange not all businesses will exchange exactly the same messages.

Figure 1 shows a typical message exchange pattern starting with a request for quote and ending with a remittance advice. Obviously purchase order change and purchase order change acknowledgement are optional. In fact depending on the business model any of these could be optional. For example, receiving goods and verifying that they reconcile against the original purchase order can be sufficient for a company to make a payment. In this case the purchase order and the remittance advice may be the only documents exchanged. A large number of companies take orders electronically, but due to the complexity, handle changes by phone or email. Whether an order can be cancelled may depend on where it is in a company's process. If it has already shipped it has to be returned etc.

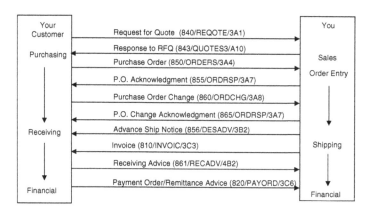

Fig. 1. Standard message exchange showing three message formats (ANSI X12, EDIFACT, RosettaNet)

1.3 Different Business Models

There are many different business models that can exist between companies, which can require greater or lesser communication between them. As well as the traditional purchase order process shown in figure 1 there are many other models such as *consigned inventory* or *call off*. *Consigned inventory* is where the supplier stores inventory on the customer's site and they only pay for it when they use it. In *call off* the customer gives the supplier forecasts that they build to, for different periods. The customer can then call off inventory, as they require it. Partial liability for part of the forecast is usual and payment is made on delivery. New processes arise all of the time, often as a result of process reengineering [18], which doesn't change the document formats, but does change which documents are used.

2 Predictions for 2010

Orchestration systems will manage all inter-system communication. Enterprise Business Integrators will have a workbench available to them that will be able to analyze a new system or business that requires integration. It will automatically reconfigure the orchestration system to connect the new system to the existing systems in the appropriate manner. The workbench will be capable of identifying new suppliers and systems on the Internet using agents. It will be able to compare them against its current configuration and recommend changes.

2.1 Orchestration Systems

Orchestration systems are available today such as Microsoft's Biztalk [20]. They manage inter-system communication and security. Adapters are available for all major applications. Documents can be received and transmitted using many technologies including FTP, HTTP, SMTP, EDI, etc. and are readily extensible. Process control languages such as XLANG [21] or BPEL [22] are used to control how a document is routed from one system to another and perform any transformation that is needed. They are able to support standards such as SOAP [23] and WSDL [24] and their support for these and other Internet standards will improve the ease of connecting systems together. However just as important, they will continue to support all of the older ways of inter-system communication providing a bridge between the future and the past. No matter how successful web services become, orchestration systems will still be needed in 2010.

2.2 EBI Workbench

The key technology that will enable the EBI workbench will be Ontologies. *"An ontology provides a vocabulary of terms and relations with which to model the domain. Ontologies are developed to provide machine-processable semantics of information sources that can be communicated between different agents (Software and Human)"* [19]. There are three different types of ontologies in figure 2.

The System ontology will describe the areas of functionality a system has and the document types and formats that the functionality expects. This would allow a workbench to determine whether the new system can communicate with its other systems and reconfigure the orchestration system to include the new system. It would do this by matching the appropriate inputs and outputs based on the existing systems ontologies and the ontology of the new system.

The business ontology will describe the capabilities of the business and the document formats it expects and supplies. Security and connection capabilities will be described. It will contain critical metrics associated with the business capabilities. This will allow the workbench to reconfigure the orchestration system to connect to a new supplier. Agents will be able to find new suppliers on the web who have the desired level of performance based on the metrics contained in the ontology.

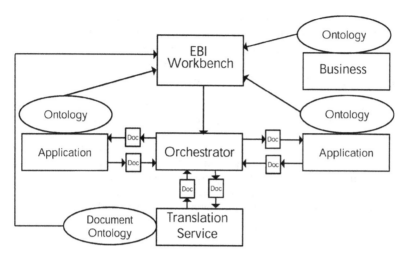

Fig. 2. EBI Workbench and it's Interaction with the Orchestration System

The document ontology will contain a definitive list of business document types and also the web services that will convert between one document type and another.

This will allow the workbench to infer a mapping from one document to another via a third document format.

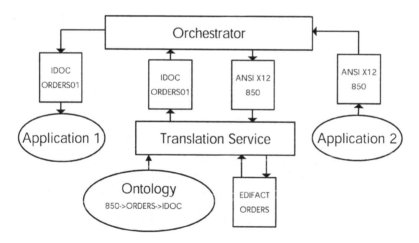

Fig. 3. Document Mediation

Figure 3 shows an example of ontology based mediation. Application 2 can supply an order in the ANSI X12 850 format. Application 1 receives orders in SAP's IDOC ORDERS01 format. The orchestrator receives the order and sends it to the translator. The translator does not know of a service that will translate from 850 to ORDERS01. It then infers using the document ontology that it can use a service to translate the 850 into an EDIFACT ORDERS format and it can use another service to translate the ORDERS format into the ORDERS01 format. It then gives the ORDERS01 document back to the orchestrator, which then passes it to application 1.

4 SCOR a Potential Ontology Base

In order to create an ontology to describe systems or businesses we must first find a vocabulary and a set of relationships that are generally agreed on. The Supply Chain Operations Reference-model (SCOR) is the industry-standard supply chain management framework promoted by the Supply Chain Council (SCC) [25], which is an independent, not-for-profit, global corporation with membership open to all companies and organizations. It has over 800 company members including: practitioners, technology providers, consultants, academics, and governments. They come from a wide range of industries and include: manufacturers, distributors and retailers. SCOR was used by HP and Compaq to plan their merger.

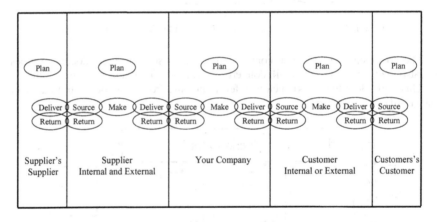

Fig. 4. SCOR Supply Chain

SCOR allows an enterprise to map processes across it's supply chain from it's supplier's supplier to its customer's customer. At SCOR level 1 there are five processes: *plan, source, make, deliver* and *return. Source* is all of the processes carried out with a supplier. *Deliver* covers all of the process carried out with a customer. *Make* is all of the processes involved in manufacturing. These three activities are linked across the supply chain. An end customer would only have a *source* process, whereas an end supplier would only have a *deliver* process. Distributors have only *source* and *deliver* processes, whereas manufacturers have *source, make* and *deliver* processes. *Return* covers all processes involved in the returning of goods. *Plan* processes happen at the intersection of any of the other four processes as shown in figure 4. This is the highest level of the SCOR model. It defines, at this level, metrics that allow the supply chain to be compared against best in class. These are: delivery performance, fill rates, perfect order fulfillment, supply chain response time, production flexibility, cost of goods sold, total supply chain management costs, value added productivity, warranty/returns processing costs, cash-to-cash cycle time, inventory days of supply, and asset turns.

At level 2 the processes are further broken down into 30 process categories. For example source is broken down into: *source stocked product*, *source make-to-order product*, and *source engineer to order product*. Each of these processes has a set of metrics associated with them allowing a business to evaluate different links in the supply chain. The process *source stocked product* for example has the following metrics: percentage of orders/lines processed complete, total source cycle time to execution, time and cost related to expediting the sourcing process of procurement, delivery, receiving and transfer, product acquisition costs, and inventory DOS.

Level 3, shown in figure 5, breaks the process down further and adds the expected inputs and outputs along with metrics for each process. This level is very useful for an ontology because it uniquely identifies the processes as in: S1.3 is *Schedule Product Deliveries*.

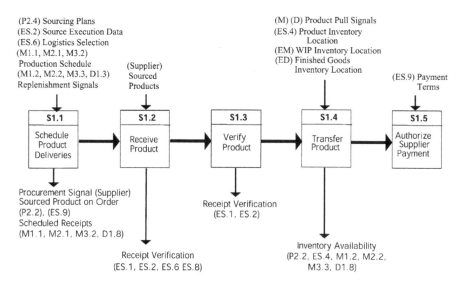

Fig. 5. S1: Source Stocked Product

It also identifies the processes, which supply the inputs and outputs. This allows a thread diagram to be constructed for the whole supply chain. A simple one is shown in figure 6. The diagram shows how SCOR can be used to map the uniquely identified processes and their interactions. In the diagram *Company 1* is purchasing stocked product from *Company 2*. It shows the documents that flow between processes. On a full model there would be one *swim lane* for each internal organization/department and each customer/supplier in the supply chain.

Some members of the Supply Chain Council have been trying to map RosettaNet PIP processes to the SCOR model. This is a very interesting piece of work, which will be useful for companies standardizing on RosettaNet. There also needs to be a way to integrate SCOR with more than RosettaNet to make it suitable for an ontology to describe systems and businesses for the purpose of integration.

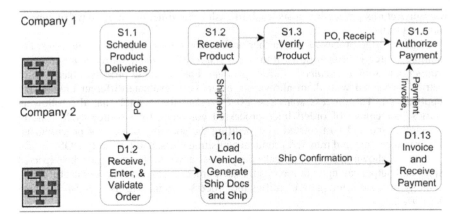

Fig. 6. Simple Thread Diagram

SCOR is, I believe a good candidate for a systems ontology. It has standard descriptions of management processes along with a name for each input and output. Most importantly it uniquely identifies processes and has a framework of relationships among those processes. The description of the capabilities of a business or system and the document formats they have available for each of the input/output types could be structured in an appropriate format such as RDF [26]. This would allow a workbench in conjunction with a document translation ontology to automatically configure an orchestration system to connect to the new business or system. Using figure 6 as an example the workbench would look at *Company 2's* D1.2 process and see that the order format that it required was an EDIFACT ORDERS format and that its equivalent of the order message is produced from it's S1.1 process. It would see the S1.1 process produced an ANSI X12 850 document and use the document translation ontology to determine the mapping.

SCOR has standard metrics to measure process performance. These are used by industry to compare suppliers and evaluate the performance of the supply chain. Structured in an appropriate format such as RDF it would enable web crawlers to look for potential suppliers who fitted or exceeded a business's objectives.

Businesses, once they were aware that web crawlers were looking for them, would use their metrics as an advertisement mechanism, in the same way as today they add keywords to their web pages to increase their prominence in web searches. Certification of metrics will be key. Some SCC members have started to publish their metrics to other SCC members.

SCOR covers all customer interactions, from order through to paid invoice. It covers all product (physical material and service) transactions, from the supplier's supplier through to the customer's customer, including: equipment, supplies, spare parts, bulk products, software, etc. Also covered are all market interactions, from the understanding of aggregate demand to the fulfillment of each order. SCOR does not describe: Sales and Marketing (Demand Generation), Research and Technology Development, Product Development, some elements of post-delivery support, Banking, Education, etc. But it is very widely used by industry and military for evaluating supply chain options and software selection.

4 Going Forward

This paper has identified SCOR as a good starting point for an EBI ontology. One difficulty, is that there is no standardized list of business documents. A new ontology needs to be created, that has a list of standardized document names as explicit as the SCOR process identifiers. It would enable the association of a generic document name with a process, which then can be cross referenced to the equivalent document from a particular standards body, such as: X12, EDIFACT or RosettaNet. The EIDX Business Process Cross-Reference [27], the ebXML catalogue of common business process [28], or SAP's Interface Repository [29], may be a good starting point. As SCOR does not cover all endeavors there will be more than one ontology. The ability to mediate between these will be key.

These ontologies, created and populated, would allow a revolution in the way that Enterprise Business Integration is being done today, by simplifying one of the hardest aspects, which is analysis of the system/business to be integrated and the existing systems. It is my belief that from an EBI point of view, we will be able to accommodate the *unreasonable man* better in 2010 A.D. by shifting from standardization to mediation.

Reference

1. S Stephens. Supply Chain Operations Reference Model Version 5.0: A new Tool to Improve Supply Chain Efficiency and Achieve Best Practice. *Information Systems Frontiers 3:4, 471–476, 2001*
2. T. Berners-Lee, J. Hendler, and O. Lassila. The Semantic Web. *Scientific American May 2001.*
3. Worldwide Enterprise Business Integration Spending Forecast and Analysis 2002–2006, *Aberdeen Group.*
4. Web Services Activity Statement, *http://www.w3.org/2002/ws/Activity*
5. HTTP, *http://www.w3.org/Protocols/rfc2616/rfc2616.html*
6. CORBA, *http://www.omg.org/gettingstarted/corbafaq.htm*
7. DCOM, *http://www.microsoft.com/com/*
8. .NET, *http://www.microsoft.com/net/*
9. ASN.1, *http://www.itu.int/ITU-T/studygroups/com17/languages/*
10. Extensible Markup Language (XML), *http://www.w3.org/XML/*
11. ANSI X12, *http://www.eidx.org/publications/document_index.html#ASCX12*
12. EDIFACT, *http://www.unece.org/trade/untdid/*
13. RosettaNet, *http://www.rosettanet.org*
14. UBL, *http://www.oasis-open.org/committees/tc_home.php?wg_abbrev=ubl*
15. CommercOne, *http://www.commerceone.com/*
16. Ariba, http://www.ariba.com/
17. OAGIS, *http://www.openapplications.org/*
18. M. Hammer, and J. Champy, Reengineering The Corporation,1993, *Harper Business.*
19. D. Fensel, Ontologies: A Silver Bullet for Knowledge Management and Electronic Commerce, 2001, *Springer-Verlag*
20. http://www.microsoft.com/biztalk/
21. S. Thatte, XLANG Web Services for Business Process Design, 2001, *Microsoft.*
22. T. Andrews, F. Curbera, H. Dholakia, Y. Goland, J. Klein, F Leymann, K Liu, D Roller, D Smith, S. Thatte, I. Trickovic, and S. Weerawarana, Business Process Execution Language for Web Services., 2003
23. M. Gudgin, M. Hadley, N. Mendelsohn, J. Moreau, and H.F. Nielsen, SOAP Version 1.2,W3C Recommendation, June 2003.

10   P. Croke

24. R. Chinnici, M. Gudgin, J. Moreau, and S. Weerawarana, Web Services Description Language (WSDL) Version 1.2, W3C Working Draft.
25. http://www.supply-chain.org
26. O. Lassila and R. R. Swick, Resource Description Framework (RDF) Model and Syntax Specification, W3C Recommendation, February 1999.
27. http://www.eidx.org/publications/xref_process.html
28. P. Levine, M. McLure, N. Sharma, D. Welsh, J. Clark, D. Connelly, C. Fineman, S. de Jong, B. Hayes, R. Read, W. McCarthy, M. Rowell, N. Sharma, and J. Loveridge, ebXML Catalog of Common Business Processes v1.0, OASIS Technical Report, May 2001.
29. http://ifr.sap.com/catalog/query.asp

Old Ontology Wine in New Semantic Bags, and Other Scalability Issues

Robert Meersman

VUB STARLab

The Web is evolving from an "eyeball web", where cognitive processing of presented information is done by humans, into an "agent web" where such processing is fully automated, leaving man-system interaction only to a presentation layer. (In fact we could posit that the so-called Semantic Web would –strictly speaking of course– be wholly unnecessary if its users would be "merely" humans.)

Re-tooling the information on the entire internet with machine-processable semantics is a huge undertaking that needs to be structured and phased, or rather, technology and methodology need to be invented such that this re-tooling process structures and phases itself to an as large extent as possible. Perhaps paradoxically, the latter approach does not require sophisticated and "intelligent" systems (these likely would fail due to the size and complexity of the information represented on the Web) but rather the adoption of simple, easily standardizable, and scalable principles, techniques, and description formalisms or languages. Examples in favor of this simplicity effect are the wide acceptance of XML, XML-Schema, and the core concepts of DAML. Examples to its contraposition are the relatively low penetration of OIL (OWL), and many ad-hoc "smart" modeling formalisms including a lot of work on e.g. upper ontologies. It goes without saying that this does not imply a negative value assessment on those techniques and research, often quite to the contrary.

At this stage serious progress is needed on theories and foundations of applied semantics, especially of evolving knowledge bases, and therefore also of the workflow aspects of establishing large, distributed ontologies as sharable, perhaps even standardized, resources on the internet.

Next, methodologies with proven properties of scalability and of reusability of existing resources are required. Scalability may be defined as the "gracious degradation" of a solution's (technique, method, system) complexity and/or performance with problem size. Ideally, the solution's complexity will even be independent (or almost so) of problem size… Storage components of a DBMS are a very good example of such near-optimal behavior, the inferencing mechanisms of many AI expert systems often exhibit the opposite; yet both are needed in the handling of large ontologies, and there will in general be a trade-off to be considered.

In this talk I will try to present, aside from certain generic definitions and terminology related to formal ontologies and formal semantics of information systems, some "lessons learned" from database design that may, or may not, carry over to ontology engineering today. Data models indeed share many fundamental properties and techniques with ontologies, but there are also many equally fundamental differences.

Where possible and appropriate I will attempt to illustrate some of these multiple aspects using the DOGMA paradigm and methodology [Development of Ontology-Guided Mediation and Applications] under research at VUB STARLab, and using some of the supporting tools under development for this purpose, as well as their ap-

C. Bussler et al. (Eds.): WES 2003, LNCS 3095, pp. 11–12, 2004.

plications. An important application area is provided by ontologies for *regulatory compliance*, and a number of projects are underway in this domain. In DOGMA we e.g. exploit a rigorous decomposition of an ontology into on the one hand a *lexon base* of elementary facts (grouped into sets corresponding to possible worlds) and on the other hand a layer of (multiple) *commitments* to derive a more structured (and scalable) approach for designing and engineering them. This approach to some extent also allows reuse of, and hence a growth path from, existing database schemas.

Meta-workflows and ESP: A Framework for Coordination, Exception Handling and Adaptability in Workflow Systems*

Akhil Kumar [1], Jacques Wainer [2], and Zuopeng Zhang [1]

[1] 509 BAB Building, Penn State University, University Park, PA 16802, USA
{akhilkumar,zoz102}@psu.edu
[2] Institute of Computing, State University of Campinas
Campinas, 13083-970, Sao Paulo, Brazil
wainer@ic.unicamp.br

Abstract. This paper describes a framework for flexible workflows based on events, states, and a new kind of process called a meta-workflow. Meta-workflows have five kinds of meta-activities and facilitate control over other workflows. We describe the framework and illustrate it with examples to show its features. The paper gives an architecture for incorporating it into existing workflows and also provides a formal semantics of execution. This framework can be used in web services, supply chains, and inter-organizational applications where coordination requirements are complex, and flexible and adaptable workflows are needed. It is also useful for handling, not just failure recovery, but also other kinds of exception situations, which arise frequently in web-based applications.

1 Introduction

One important aspect of Web services and supply chains [15,17,6] is their nature as a workflow that spans multiple organizations. They require coordination of both the data flow and control flow across multiple organizations. A major problem that arises in workflows (both intra- and inter-organizational) is that of exceptions or special situations. Even a seemingly simple process like travel expense claim or order processing can become difficult to describe if one tries to cover all the special situations and exceptions in the description. This creates a very awkward process description that is hard to read and understand, and also error-prone. Therefore, most workflow systems are able to capture the simpler form of a process and tend to collapse when variations are introduced.

Exceptions arise frequently in workflow systems [2,4,13,18]. A simple travel expense claim processing example involves steps like submit expense claim, review by secretary, approval by manager, and payment. However, even in such a simple example, numerous exceptions can arise. For instance, the manager might wish to get

---

* This research was supported in part by a grant from the IBM Corporation through the Center for Supply Chain Research at Penn State University.

C. Bussler et al. (Eds.): WES 2003, LNCS 3095, pp. 13–27, 2004.

some clarifications from the employee, or may disapprove certain expenses. Alternatively, the manager may not be available or may have left the company, in which case some alternative must be found. Moreover, if the expense claim is not processed in a timely manner, then somebody should be notified. Exceptions such as these can be *planned* or *unplanned*. A planned exception is an abnormal situation that has been anticipated and a way for handling it has been included in the process. On the other hand, an unplanned exception is one that has not been anticipated. Picking up on the above example, a manager required to approve expenses incurred in travel to Tokyo may decide to transfer the approval responsibility to her boss or another manager who has visited Tokyo and knows about the range of typical expenses there.

In this paper, we describe a formal methodology for describing exception situations in a workflow. We consider an exception as a special situation that occurs infrequently in a workflow. The main idea is to describe a basic, main-line process first and treat abnormal and infrequent situations separately as exceptions, some anticipated and others not. Our goal is to provide support for the planned exceptions and also be able to incorporate the unplanned ones relatively easily. We introduce two new notions, meta-workflow and ESP (Event-state-process). A *meta-workflow* is a special, higher level control process that consists of five control commands: start, terminate, suspend, resume and wait. An *ESP rule* causes a meta-workflow to run when an event occurs and a workflow case is in a certain state. This may cause a meta-workflow to execute, and thus perform control operations like suspending certain workflows and starting other workflows, etc. Thus, we make a clear distinction between two types of workflows: *base-workflows* and *meta-workflows*. The base-workflow corresponds to specific tasks that must be performed. The meta-workflow is only for control purposes and consists of the control commands described above. In general, when it is not qualified the term workflow refers to a base-workflow.

We foresee several advantages of this approach. First, it leads to *modularity*. The basic workflow description is kept very simple while variations in the basic process are described separately. Thus, there can be one main base-workflow and n other supporting base-workflows describing the variations of the main base-workflow. Second, it provides *extensibility* and *adaptability*. It is not possible to determine all possible exceptions that may arise for a process, even a deceptively simple one like the ones described above. Thus, it is possible to add new ESP rules and corresponding base and meta-workflows when a new situation arises that was not anticipated. Finally, the simplicity of the approach helps in minimizing errors, while at the same time making it easier to describe complex situations.

Consequently, this framework constitutes a new methodology for workflow modeling which has applications in various kinds of web services. The organization of this paper is as follows. Section 2 gives a formal description of our framework along with semantics. Then section 3 illustrates the framework with examples designed in the context of WQL (Workflow Query Language). Next, Section 4 discusses an architecture, while in Section 5 we present some discussion of this approach in the context of related work. Finally, Section 6 concludes this paper.

2 Formal Description

The ESP framework consists of events, states and processes. At the outset it is important to keep in mind the distinction between process classes (or templates) and process instances. A process definition, say "fabrication of a car" is a process class (or a template), composed of three activities, A, B, and C, to be performed in sequence. Activities A, B and C also belong to activity classes or templates. Both the fabrication process and the three activities are generic.

A particular instance of "fabrication of a car," say the car with the VIN XYZ345, is a case, or an instance of a process. Case XYZ345 (assuming the VIN is used as the case id) may be executing activity B; i.e., the activity B (for case XYZ345) is in the executing state. In the ESP framework, as we will see later, multiple processes may be invoked with the same case number. Thus, a process instance is initiated and assigned a case number, and the same case number is used as a reference when other process templates are invoked. For example, suppose the fabrication of car XYZ345 is to be canceled while B is the current activity. Hence, B must be stopped, and a new cancellation activity, say "returning the reusable parts to the inventory" might start, and other activities may follow. The case is still identified by id XYZ345, but, of course, it is no longer an instance of the fabrication process, but an instance of the cancellation process. We assume that cases are identified by a case-id. Furthermore, we will assume the standard states for an activity instance:

- **not-ready**: if some of its prerequisite activities are not done.

- **ready**: if all its prerequisites are done, but the activity has not yet started.

- **executing**: if the activity is executing.

- **suspended**: if the activity is suspended.

- **done**: if the execution of the activity has terminated.

In the ESP framework, a single case may be executing in multiple concurrent workflows. In the example above, when the cancel event was raised, the "fabrication of a car" workflow was **suspended**, and a new workflow, "dealing with cancellation," was started. But, in general, it may also be the case that more than one workflow is **executing** for the same case.

In this framework, there are potentially multiple concurrent workflows. A workflow instance is **active** if any of its activities is not **done**; otherwise, the workflow case is **inactive**, which may mean that the case has already ended (all activities are **done**) or has not yet started.

An ESP framework (see Figure 1) has two components, the workflow **description** part and the **activation rules**. The description component is of the form:

```
base-workflow-id1: workflow-definition1

base-workflow-id2: workflow-definition2

meta-workflow-id3: workflow definition3

meta-workflow-id4: workflow definition4
```

The description component associates a workflow name to a process definition, so that the activation rules can only refer to the name of the process. It also associates a

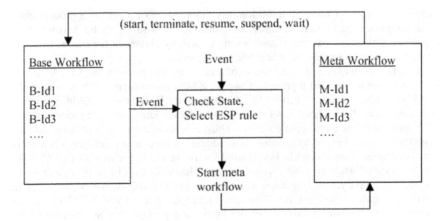

Fig. 1. Overview of meta-workflow and ESP framework

name to the process definitions of the meta-workflows. The workflow definition language is not central to this paper, but a language called WQL will be described briefly in the context of some examples in the next section.

The activation rules contain two parts. The first part, for **base-workflows,** just lists the workflow ids of these workflows in the form:

```
base-workflow-id1

base-workflow-id2
```

The second part, for the **exception rules**, is of the form:

```
base-workflow-id1, event-class1, state1 → meta-
workflow-id1

base-workflow-id2, event-class2, state2 → meta-
workflow-id2
```

The base-workflow component of the activation rules just lists all standard workflows, the ones that behave as expected: upon receiving some signal to start the workflow, a new case is created and the first activity in the workflow becomes **ready**.

The basic semantics of an exception rule is the following. If an event of *event-class* is received for a case, and the base-workflow given by the *base-workflow-id* is active and is in the state given in the rule, then the corresponding meta-workflow is started for the case.

- **base-workflow-id** is the id or name of a workflow. We assume that all workflows have a unique identification.

- **event-class** is the name of a type of event. Events are global, atomic, and contain an attribute that associates the event to a case. Thus, the situation in which case XYZ345 must be canceled is signaled by generating an event of class *cancellation* with attribute XYZ345.

- **state** is a description of which activities of the base-workflow are ready, active, and done. We will discuss how states are represented below.

- **meta-workflow** is a workflow which may make use of five "meta-activities" (i.e., activities that control the base-workflow). We will discuss the meta-activities below. Also the meta-workflow has a corresponding **meta-workflow-id**.

The name **ESP** comes from a previous version of the semantics above, but it stands for **E**(events) **S**(states) **P**(process), a deviation from the more common Event, Condition, Action (ECA) rules [16].

2.1 Meta-activities

What distinguishes a meta-workflow from a base-workflow is that a meta-workflow only has five special activities, called **meta-activities** that control and operate the base-workflows. These meta-activities are:

- **start(wf, [c])**: which activates the base-workflow wf and associates it with active case c.

- **terminate(wf, [c])**: which terminates the base-workflow wf with active case c. The terminate meta-activity suspends all executing activities, and places the base-workflow in the inactive state.

- **suspend(wf, [A], [c])**: suspends the activity A in the base-workflow wf *(for the current case)*. If activity A is not executing, it has no effect. Finally, if the argument A is missing, all executing activities are suspended.

- **resume(wf, [A], [c])**: resumes activity A in the base-workflow wf. If A is not suspended in the base-workflow, the meta-activity has no effect. If argument A is missing, then the workflow resumes where it was suspended.

- **wait(wf, [A], [c])**: this activity does nothing except wait for the activity A in the base-workflow wf to be done. If A is already done, continue(A) does nothing. If argument A is missing, then the wait persists until the workflow itself is terminated.

Note that the case argument c may be omitted from start and terminate if it is obvious from the context. Next we discuss the states of various activities and their semantics.

2.2 State Representation in ESP Framework

We will define a state representation in ESP as a logical formula which may refer to:

- Activities or sets of activities.

- The set Done, which represents the set of activities in the **done** state.

- The set Executing, which represents the set of activities in the **executing** state.

- The set Ready, which represents the activities in the **ready** state.

- Equalities or inequalities that refer to case data.

Thus the formula:

```
(A ⊂ Done ∧ {B,C} ⊆ Executing ∧ Cost ≤ 300,000) ∨
({A,B,C} ⊆ Done ∧ ¬ D ⊂ Done)
```

represents the situation in which A is **done,** B and C are still **executing** and the case data *cost* is less than or equal to $300,000, or A, B, and C are all **done** but D is not.

2.3 Semantics

We assume that set C gives the case-ids of the current cases. For each $c \in C$ there is at least one workflow *wf* that is active for case c (there may be more than one), denoted as *wf* $\in$ active(c). For each active workflow *wf* for c there is a set of activities of *wf* which are in the **done** state, denoted by $D(c,wf)$, and a set of activities which are **executing**, denoted by $E(c,wf)$.

The semantics for the execution of base-workflows is as follows (where *wf* is a workflow-id, and c a case id):

> **if** the engine receives the event start(*wf, c*) from the meta-workflow module
>
> **then**
>> insert c into C, if not present
>>
>> insert *wf* in active(c)
>>
>> place the first activity of *wf* for case c in the **ready** state.
>
> **end if**

For ESP rules of the form wf_i: event$_i$, state$_i$ → mwf_i, the operational semantics is the following. Upon receiving an event $e(c)$ where e is an event class and c a case-id:

> **for** each rule i **do**
>> **if** $wf_i \in$ active(c) and $e =$ event$_i$ and state of wf_i for case c satisfies
>>
>>> the expression for state$_i$ **then**
>>>
>>>> pass ids mwf_i and c *to the meta-workflow module*
>>
>> **end if**
>
> **end for**

The above definition implies that:

1. more than one rule can capture an event,

2. there are cases in which the event is not captured by any rule

3. having the event captured by a rule has, by itself, no impact on the base-workflow wf_i that was active when the event was received.

If an event is captured by more than one rule, then conflict resolution is required to select one rule. For simplicity, we propose to associate priorities with rules to resolve conflicts. If an event is not captured it is ignored. Any changes in the base-workflow state can only be accomplished by the execution of meta-activities in a meta-workflow.

Let us discuss formally the effects of the meta-activities. While running a meta-workflow the meta-workflow module sends meta-activities as instructions to the main workflow engine. In general, the parameters for a meta-activity X are a workflow *wf* and a case *c*. The semantics for the action taken by the workflow engine on each X are as follows.

if X = **start(*wf, c*) then**

  insert *c* into *C*, if not present

  insert *wf* in active(*c*)

  place the first activity of *wf* for case *c* in the **ready** state

else if X = **terminate(*wf, c*) then**

  suspend all activities executing in *wf* for *c*

  remove *wf* from active(*c*)

else if X = **suspend(*wf, [A], c*) then**

  if A is specified **then** suspend activity A;

  if A is unspecified **then** suspend all executing activities of *wf*

else if X = **resume(*wf, [A,], c*) then**

  if **A** is specified, and **A** is in **suspended** state, **then** resume *A*

  if **A** is unspecified, **then** **resume** all suspended activities

else if X = **wait (*wf, A, c*) then**

  if A is executing in *wf* for *c* **then**

    wait until A is **done** in *wf* for *c*

  end if

end if

3 WQM Model and Examples

We use the WQM (workflow query model) to illustrate the applications and implementations of meta-workflows ([5]). In WQM, a workflow process is described in terms of nine basic primitives: *start, task, end, split-choice, join-choice, split-parallel, join-parallel, start-while_do,* and *end-while_do* (see [5] for more details). This simple language can be used to model complex inter-organizational workflow processes to create web services such as the one we discuss next.

Consider a base-workflow, wf_1, as showed in Figure 2 [5] for ordering a laptop. This workflow includes several tasks or activities such as entering the order, obtaining

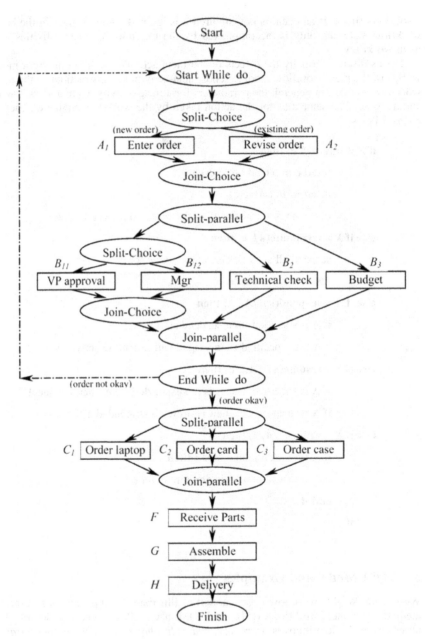

Fig. 2. An example base-workflow, wf_1, using WQM model [5]

multiple approvals, placing separate orders for components of the laptop, receiving the parts, assembly and delivery. The split-choice node allows one path to be taken based on a condition (e.g., new/old order). The split-parallel allows multiple activities to proceed in parallel. The While-do construct is used to expression repetition. The various activities of this workflow process have been labeled (A, B_{11}, B_{12}, etc.) in the figure for ease of reference.

In this section, we use three event examples to illustrate our framework. Each event is also associated with an active case c.

3.1 Example 1: Late Notification After Activity F

If the parts arrive late, the assembly line, the final customers, and the suppliers will be notified before the next activity G is executed. We name this workflow process in Figure 3 as wf_2, which is another base-workflow that will be triggered by meta-workflow mwf_1 if $event_1$ = Late(F) is received by the system and wf_1 is in $state_1$. States, meta-workflows and exception rules for this event are shown in Table 1.

The event Late(F) will be captured by the ESP rule to trigger the execution of meta-workflow mwf_1 or mwf_2, depending upon the state of the workflow wf_1. If the activity F, "receive parts", is already **done**, the meta-workflow mwf_2 will first suspend the next activity in base-workflow wf_1 and then start the base-workflow wf_2. If the activity F is still executing when $event_1$ is received by the system, then the meta-workflow mwf_2 will first wait for the completion of activity F, and then perform the same meta-activities as those in the previous scenario. After the base-workflow wf_2 is finished, another ESP rule will capture event Done(wf_2), start workflow mwf_3, and thus, resume the next activity in base-workflow wf_1.

Notice that the same event can be captured by multiple ESP rules under different states. When the activity F is in the state of **executing** or **done**, $event_1$ will be captured by different ESP rules and trigger meta-workflow mwf_1, or mwf_2 respectively. For now we are assuming that once an event is captured and processed it will be disabled to prevent repetitive firing of meta rules.

3.2 Example 2: Process Abortion

After the activity "enter order" is done, one may want to abort the whole process if some relevant order details are missing or incorrectly provided by the customer. For example, the customer may have entered the wrong credit card information. Then, to abort the entire process, the workflow process in Figure 4 can be executed to undo the partial effect of the activity "enter order" and remove the information that has already been recorded by the database. We name this workflow process as wf_4 and it is triggered by meta-workflow mwf_6 if $event_5$ = MissingInfo(A_1) is received by the system and activity A1 is done. The meta-workflow will first suspend the base-workflow wf_1, then activate and run the reverse workflow wf_4, and finally terminate the original base-workflow wf_1. The states, meta-workflows, and exception rules for this event are shown in Table 2.

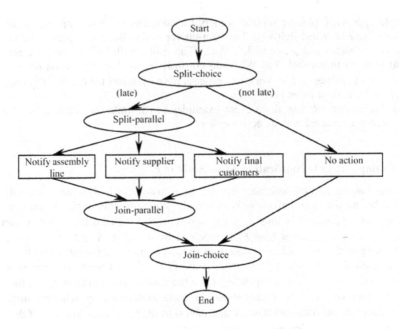

Fig. 3. A base-workflow, wf_2, for late notification

Table 1. Formal definition of late notification

Events	event$_1$ = Late(F)
	event$_2$ = Done(wf_2)
Meta-workflows	mwf_1: suspend(wf_1, G) → start(wf_2)
	mwf_2: wait(wf_1, F) → suspend(wf_1, G) → start(wf_2)
	mwf_3: resume(wf_2, G)
ESP rules	wf_1: Late(F): $F \in Done \rightarrow mwf_1$
	wf_1: Late(F): $F \in Executing \rightarrow mwf_2$
	wf_2: Done(wf_2) → mwf_3

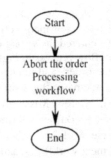

Fig. 4. A reverse workflow, wf_4, for process abortion

Table 2. Formal definition of process abortion

Events	$event_5 = \text{MissingInfo}(A_1)$
Meta-workflows	mwf_6: suspend(wf_1) → start(wf_4) → terminate(wf_1)
ESP rules	wf_1: MissingInfo(A_1): $A_1 \in Done \rightarrow mwf_6$

3.3   Example 3: Additional Budget Approval

Now we turn to a process that will perform an "additional budget check" after the regular "budget check" activity is done, if the budget for this order is found to be greater than \$3000 since that is the maximum allowed by company policy for a laptop. In this case, the VP of Finance and the President must approve the request as an exception since it exceeds the company limit. We name this sub-workflow process shown in Figure 5 as wf_5. It will be triggered by meta-workflow mwf_9 if $event_4 = \text{Done}(B_3)$ is received by the system and if the budget amount is greater than \$3000. The ESP rule will trigger a meta-workflow to suspend wf_1 and start wf_5 to perform the additional reviews. After the additional approvals finish, the event Done(wf_5) will trigger the meta-workflow mwf_{10} to resume the original base-workflow wf_1. The states, meta-workflows, and exception rules for this event are shown in Table 3.

Fig. 5. A base-workflow, wf_5, for additional budget approval

Table 3. Formal definition of additional budget check

Events	$event_6 = \text{Done}(B_3)$
	$event_7 = \text{Done}(wf_5)$
Meta-workflows	mwf_9: suspend(wf_1) → start(wf_5)
	mwf_{10}: resume(wf_2)
ESP rules	wf_1: Done(B_3): Budget_amount > \$3000 → mwf_9
	wf_5: Done(wf_5) → mwf_{10}

In this section, we have seen how different types of special situations can be described using our framework. The states can be based both on the status of various tasks and also on the case data.

4 Architecture

The architecture for incorporating this framework is shown in Figure 6. In this architecture we extend an existing workflow enactment service with two modules: an event support module and a meta-workflow module. This architecture was developed to make the most use of existing workflow engine architectures. The component on the left in Figure 6 is a standard workflow engine; the components on the right are new additions. Thus, it is conceivable that a legacy workflow may run entirely in the workflow engine.

The ESP support module allows users to specify rules. It receives the events both from the (standard) workflow engine, and from external sources and case data from the workflow engine and selects a rule and a corresponding meta-workflow to run. Then, it passes the meta-workflow and case information to the meta-workflow execution support module. The latter manages the execution of the meta-activities by sending them to the main workflow execution module.

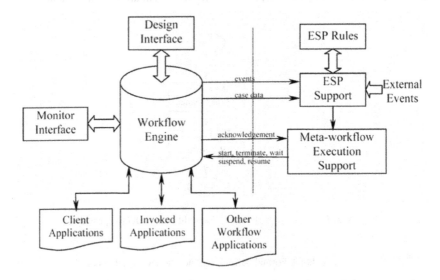

Fig. 6. An architecture for the ESP framework

To support this architecture, the following modifications are proposed in an existing workflow engine. After each activity finishes, the main workflow engine must:

- suspend the current path of the workflow process.
- send new events to the ESP module.

- send case data to the ESP module.

- wait for instructions from the ESP module.

The workflow engine also needs to be able to receive and execute the five new meta-activities. Most workflow engines provide APIs that would make such modifications reasonably easy to implement.

5 Discussion and Related Work

Support for events can play an important role in modeling of inter-organizational workflows because events offer a convenient mechanism for coordination. The need for such support has been noted elsewhere also (see [10] for example). This paper has proposed to integrate events with meta-workflows to create a powerful coordination and control mechanism. The usefulness of this approach was illustrated through various examples.   Coordination requirements can be quite complex in inter-organizational workflows [1,8,9]. Our approach can have considerable value in the context of Web services [17] and supply chain applications [6]. In Web services, meta-workflows can be used to facilitate inter-operation between multiple related services (e.g., airlines, hotel and car rental reservations) that must be integrated. In supply chain applications, exceptions like missed deliveries, stock-outs, etc. arise quite often. Here our framework can assist in reacting to such new events in a systematic manner and improving the level of collaborative information sharing between partners.   Recently there has also been interest in adaptive workflows [12,11].   This framework can be used for adapting workflows dynamically as illustrated by the sub-workflow substitution and insertion examples earlier.

A prototype event engine, called EVE, for implementing event-driven execution of distributed workflows has been presented in [7].   However, the most relevant related work to ours is the event-based inter-process communication in the context of OPERA [10]. The mechanism described in [10] allows processes to communicate by means of event based control connectors (ECCs). An ECC is associated with an event and, upon occurrence of the event, if a condition is true then another process or an activity can be invoked. Our approach is similar in spirit to this work; however, we make a clear distinction between workflows and meta-workflows, which is lacking in [10]. This separation allows for a more systematic and flexible methodology for process design, and nicer semantics. Thus, meta-workflows serve as a useful modeling construct for controlling multiple workflows and dealing with various kinds of special (exception) situations that often arise, including failure handling, recovery, etc.

Research on exceptions in workflows is still limited. In [18], Strong and Miller define exceptions as "cases that computer systems cannot correctly process without manual intervention." Based on a field study they make several recommendations, such as the need for more efficient exception handling routines and better support for people who have to fix exceptions. Borgida and Murata [2] describe exceptions as violations of constraints and apply ideas from exception handling in programming languages. They treat an exception as an object belonging to a class that can have attributes. Their class structure is similar to a taxonomy. Another taxonomy-based approach to handling exceptions by Klein and Dellarocas is presented in [13]. They define an exception as "any departure from a process that achieves the process goals completely and with maximum efficiency." The authors propose to create a taxonomy based on the type of exception and have corresponding strategies to be applied once an exception can be classified. The approach in [4] is based on ECA style rules. These rules are bound to the workflow for exception handling at different levels of scope.

As noted in [10] also, the event-based framework has some similarities with the ECA style rules [16]. Clearly, both approaches are based on rules and events. However, the application environments and semantics are different. ECA rules are used in active databases in the context of changing data. ECA has also been proposed in the WIDE and EVE prototype workflow systems [3,7] as a means to describe the coordination requirements in a workflow itself and to handle simple exceptions. The major difference is that the action in ECA corresponds to a sub-transaction (such as updates to the database). In workflows, instead of the action a special process called a meta-workflow is executed resulting in vastly different semantics. Therefore, an attempt to implement this framework in a database that supports ECA rules is likely to be impractical and awkward.

Finally, the ESP framework must be compared with some more standard Web Services specification languages, such as BPEL4WS. In broad terms, events in ESP correspond to faults and events in BPEL (but ESP does not distinguish between them), ESP rules correspond to fault- and event-handlers, caseId corresponds to the BPEL notion of correlation sets and properties, and so on. The main contribution of ESP is a richer language to express exceptions than BPEL's. In particular, ESP rules allow considerable flexibility to terminate and suspend executing activities (not necessarily the ones that generated the faults) and to resume execution of suspended activities as well. Furthermore, the way to express some of the kinds of exception conditions allowed in ESP would be awkward in BPEL, if at all possible.

6 Conclusions

This paper described a framework and architecture to support handling of exceptions and various kinds of special situations that arise in a workflow system. It is based on combinations of events and states that cause higher-level processes called meta-workflows to be executed. A meta-process consists of five meta-activities for controlling the behavior of and coordinating base-workflow processes. We demonstrated the usefulness of the approach and gave a preliminary architecture for integrating this approach into a current workflow system. The advantages of this framework are modularity, extensibility, adaptability and simplicity.

This work is still ongoing and we foresee further research along several lines. First, we would like to extend the current XRL (eXchangeable Routing Language) effort [1] for describing inter-organizational workflows in XML by adding this functionality to both the language and prototype. Secondly, although we gave the semantics for execution of meta-workflows themselves, rule conflicts have not been considered in this paper, and we need to address this by developing a more detailed rule execution semantics. Thirdly, support for composite events would also be a useful feature. Next, the proposed framework should be evaluated in terms of its control capabilities and performance. Finally, the architecture needs to be extended for a fully distributed environment.

References

1. W.M.P. van der Aalst and A. Kumar, "XML Based Schema Definition for Support of organizational Workflow", *Information Systems Research* Vol. 14, No. 1, 23–46, March 2003.
2. Borgida, A. and Murata, T., "Tolerating exceptions in workflows: a unified framework for data and processes", *Proceedings of WACC'99*, 59–68.
3. Casati, F., et al., "Deriving active rules for workflow management", *Proceedings 7th DEXA*, Zurich Switzerland, September 1996.
4. Li, C. and Karlapalem, "A Meta Modeling Approach to Workflow Management Systems Supporting Exception Handling", *Information Systems*, Vol.24, No.2, pp. 159–184, 1999.
5. Christophides, V., Hull, R., and A. Kumar, "Querying and Splicing of XML Workflows", *CoopIS 2001*: 386–402.
6. Curran, T. and Ladd, A., "SAP R/3: Understanding enterprise supply chain management", Prentice-Hall, 2000.
7. Geppert, A. and D. Tombros, "Event-based distributed workflow execution with EVE", *Technical Report 96.5*, University of Zurich, 1996.
8. Grefen, P., Aberer, K., Hoffner, Y.; and Ludwig, H., "CrossFlow: cross-organizational workflow management in dynamic virtual enterprises", *International Journal of Computer Systems Science and Engineering*, pp. 277–290, (15)5, CRL Publishing, London, September 2000.
9. Gronemann, B., Joeris, G., Scheil, S., Steinfort, M., and Wache, H. "Supporting cross-organizational engineering processes by distributed collaborative workflow management - The MOKASSIN approach", *Proc. of 2nd Symposium on Concurrent Mulitdisciplinary Engineering (CME'99)*, Bremen, Germany, September 1999.
10. C. Hagen, and Alonso, G, "Beyond the Black Box: Event-based Inter-Process Communication in Process Support Systems," In *Proc. of 19th ICDCS*, Austin, TX, 1999.
11. Han, Y., Sheth, A., and Bussler, C., "A Taxonomy of Adaptive Workflow Management, in: Workshop "Towards Adaptive Workflow Systems", *1998 ACM Conference on Computer Supported Cooperative Work, Seattle*, Washington, USA, November 14–18, 1998.
12. Joeris, G. and Herzog, O., "Managing Evolving Workflow Specifications", *3rd IFCIS Intl. Conf. on Cooperative Information Systems (CoopIS'98)*, 1998; pp. 310–319.
13. Klein, M. and Dellarocas, C., "A Knowledge-Based Approach to Handling Exceptions in Workflow Systems", *Computer Supported Cooperative Work (CSCW)*, Vol. 9, Number 3/4, 399–412, 2000.
14. Lazcano, A., G. Alonso, H. Schuldt, and C. Schuler, "The WISE Approach to Electronic Commerce", *Computer Systems Science & Engineering*, 15(5), pp. 345–357, 2000.
15. Lee, H. and Whang, S., "Information Sharing in Supply Chains", Stanford Graduate School of Business, Research paper 1549, 1998.
16. McCarthy, D.R. and Dayal, U., "The Architecture of an Active Database System", in *Proc. ACM SIGMOD Conf. on Management of Data*, Portland, 1989, pp. 215–224.
17. Sayal, M., Casati, F., and Shan, M., "Integrating workflow management systems with Business-to-Business Interaction standards", *HP Tech. Report*, HPL-2001-167, July 2001.
18. Strong, D. and Miller, S., "Exceptions and exception handling in computerized information processes", *ACM Trans. On Information Systems*, Vol. 13, No. 2, 1995.

A Foundational Vision of e-Services

Daniela Berardi, Diego Calvanese, Giuseppe De Giacomo
Maurizio Lenzerini, and Massimo Mecella

Dipartimento di Informatica e Sistemistica
Università di Roma "La Sapienza"
Via Salaria 113, 00198 Roma, Italy
lastname@dis.uniroma1.it

Abstract. In this paper we propose a foundational vision of e-Services, in which we distinguish between the external behavior of an e-Service as seen by clients, and the internal behavior as seen by a deployed application running the e-Service. Such behaviors are formally expressed as execution trees describing the interactions of the e-Service with its client and with other e-Services. Using these notions we formally define e-Service composition in a general way, without relying on any specific representation formalism.

1 Introduction

The spreading of network and business-to-business technologies [11] has changed the way business is performed, giving rise to the so called *virtual enterprises* and communities [7]. Companies are able to export services as semantically defined functionalities to a vast number of customers, and to cooperate by composing and integrating services over the Web. Such services, usually referred to as e-Services or Web Services, are available to users or other applications and allow them to gather data or to perform specific tasks. *Service Oriented Computing* (SOC) is a new emerging model for distributed computing that enables to build agile networks of collaborating business applications distributed within and across organizational boundaries[1].

Cooperation of e-Services poses many interesting challenges regarding, in particular, composability, synchronization, coordination, correctness verification [13]. However, in order to address such issues in an effective and well-founded way, e-Services need to be formally represented.

Up to now, research on e-Services has mainly concentrated on three issues, namely *(i)* service description and modeling, *(ii)* service discovery and *(iii)* service composition.

Composition addresses the situation when a client request cannot be satisfied by any available e-Service, whereas a *composite* e-Service, obtained by combining a *set* of available *component* e-Services, might be used. Composition involves

---

[1] cf., Service Oriented Computing Net: http://www.eusoc.net/

C. Bussler et al. (Eds.): WES 2003, LNCS 3095, pp. 28–40, 2004.
© Springer-Verlag Berlin Heidelberg 2004

two different issues: the one of *composing by synthesis* a new *e*-Service starting from available ones, thus producing a *composite e-Service specification*, and the one of enacting, i.e., instantiating and executing, the composite *e*-Service by correctly coordinating the component ones; the latter is often referred to as *orchestration* [6,10], and it is concerned with monitoring control and data flow among the involved *e*-Services, in order to guarantee the correct execution of the composite *e*-Service. In what follows, we concentrate on composition synthesis: orchestration techniques go beyond the scope of this paper.

The *DAML-S Coalition* [2] is defining a specific ontology and a related language for *e*-Services, with the aim of composing them in automatic way. In [12] the issue of service composition is addressed, in order to create composite services by re-using, specializing and extending existing ones; in [9] composition of *e*-Services is addressed by using GOLOG. In [1] a way of composing *e*-Services is presented, based on planning under uncertainty and constraint satisfaction techniques, and a request language, to be used for specifying client goals, is proposed.

All such works deal with different facets of service oriented computing, but unfortunately an overall agreed upon comprehension of what an *e*-Service is, in an abstract and general fashion, still lacking. Nevertheless, *(i)* a framework for formally representing *e*-Services, clearly defining both specification (i.e., design-time) and execution (i.e., run-time) issues, and *(ii)* a definition of *e*-Service composition and its properties, are crucial aspects for correctly addressing research on service oriented computing.

In this paper, we concentrate on these issues, and propose an abstract framework for *e*-Services, in order to provide the basis for *e*-Service representation and for formally defining the meaning of composition. Specifically, Section 2 defines the framework, which is then detailed in Sections 3 and 4 by considering *e*-Service specification and run-time issues, respectively. Section 5 describes the basic, conceptual interaction protocol between a running *e*-Service and its client. Section 6 deals with composition, in particular by formally defining such a notion in the context of the proposed framework. Finally, Section 7 concludes the paper, by pointing out future research directions.

2   General Framework

Generally speaking, an *e*-Service is a software artifact (delivered over the Internet) that interacts with its clients, which can be either human users or other *e*-Services, by directly executing certain actions and possibly interacting with other *e*-Services to delegate to them the execution of other programs. In this paper we take an abstract view of such an application and provide a conceptual description of an *e*-Service by identifying several facets, each one reflecting a particular aspect of an *e*-Service during its life time.

- The *e*-Service *schema* specifies the features of an *e*-Service, in terms of functional and non-functional requirements. Functional requirements represent *what* an *e*-Service does. All other characteristics of *e*-Services, such as those related to quality, privacy, performance, etc. constitute the non-functional requirements. In what follows, we do not deal with non-functional requirements, and hence use the term "*e*-Service schema" to denote the specification of functional requirements only.

- The *e*-Service *implementation and deployment* indicate *how* an *e*-Service is realized, in terms of software applications corresponding to the *e*-Service schema, deployed on specific platforms. Since this aspect regards the technology underlying the *e*-Service implementation, it goes beyond the scope of this paper and we do not consider it any more[2]. We have mentioned it for completeness and because it forms the basis for the following one.

- An *e*-Service *instance* is an occurrence of an *e*-Service effectively running and interacting with a client. In general, several running instances corresponding to the same *e*-Service schema exist, each one executing independently from the others.

As mentioned, the schema of an *e*-Service specifies what the *e*-Service does. From the external point of view, i.e., that of a client, the *e*-Service is seen as a black box that exhibits a certain "behavior", i.e., executes certain programs, which are represented as sequences of atomic *actions* with constraints on their invocation order. From the internal point of view, i.e., that of an application deploying an *e*-Service E and activating and running an instance of it, it is also of interest how the actions that are part of the behavior of E are effectively executed. Specifically, it is relevant to specify whether each action is executed by E itself or whether its execution is delegated to another *e*-Service with which E interacts, transparently to the client of E. To capture these two points of view we consider the *e*-Service schema as constituted by two different parts, called *external schema* and *internal schema*, respectively representing an *e*-Service from the external point of view, i.e., its *behavior*, and from the internal point of view.

In order to execute an *e*-Service, the client needs to *activate* an instance from a deployed *e*-Service: the client can then interact with the *e*-Service instance by repeatedly *choosing* an action and waiting for the fulfillment of the specific task by the *e*-Service and (possibly) the return of some information. On the basis of the information returned the client chooses the next action to invoke. In turn, the activated *e*-Service instance executes (the computation associated to) the invoked action and then is ready to execute new actions. Note that, in general, not all actions can be invoked at a given point: the possibility of invoking them depends on the previously executed ones, according to the external schema of the *e*-Service. Under certain circumstances, i.e., when the client has reached his goal, he may explicitly *end* (i.e., terminate) the *e*-Service instance. However, in principle, a given *e*-Service may need to interact with a client for an unbounded, or even infinite, number of steps, thus providing the client with a continuous service. In this case, no operation for ending the *e*-Service is ever executed.

---

[2] Similarly, recovery mechanisms are outside the scope of this paper.

For an instance e of an e-Service E, the sequence of actions that have been executed at a given point and the point reached in the computation, as seen by a client, are specified in the so-called *external view* of e. Besides that, we need to consider also the so-called *internal view* of e, which describes also which actions are executed by e itself and which ones are delegated to which other e-Service instances, in accordance with the internal schema of E.

To precisely capture the possibility that an e-Service may delegate the execution of certain actions to other e-Services, we introduce the notion of *community* of e-Services, which is formally characterized by:

- a common set of actions, called the *alphabet* of the community;
- a set of e-Services specified in terms of the common set of actions.

Hence, to join a community, an e-Service needs to export its service(s) in terms of the alphabet of the community. The added value of a community of e-Services is the fact that an e-Service of the community may delegate the execution of some or all of its actions to other instances of e-Services in the community. We call such an e-Service *composite*. If this is not the case, an e-Service is called *simple*. Simple e-Services realize offered actions directly in the software artifacts implementing them, whereas composite e-Services, when receiving requests from clients, can invoke other e-Services in order to completely fulfill the client's needs.

The community may also be used to generate (virtual) e-Services whose execution completely delegates actions to other members of the community.

In the following sections we formally describe how the e-Services of a community are specified, through the notion of e-Service schema, and how they are executed, through the notion of e-Service instance.

3 e-Service Schemas

In what follows, we go into more details about the two schemas introduced in the previous section.

3.1   External Schema

The aim of the external schema is to abstractly express the behavior of the e-Service. To this end an adequate specification formalism must be used, which allows for a finite representation of such a behavior[3]. In this paper we are not concerned with any particular specification formalism, rather we only assume that, whatever formalism is used, the external schema specifies the behavior in terms of a tree of actions, called *external execution tree*. Each node x of the tree represents the history of the sequence of interactions between the client and the e-Service executed so far. For every action a that can be executed at the point represented by x, there is a (single) successor node y_a with the edge

---

[3] Typically, finite state machines are used [8,5].

(x, y_a) labeled by a. The node y_a represents the fact that, after performing the sequence of actions leading to x, the client chooses to execute the action a, among those possible, thus getting to y_a. Therefore, each node represents a choice point at which the client makes a decision on the next action the e-Service should perform.

The root of the tree represents the fact that the client has not yet performed any interaction with the e-Service. Some nodes of the execution tree are *final*: when a node is final, and only then, the client can end the interaction. In other words, the execution of an e-Service can correctly terminate at these points[4].

Notably, an execution tree does not represent the information returned to the client, since the purpose of such information is to let the client choose the next action, and the rationale behind this choice depends entirely on the client.

Example 1. Figure 1 shows an execution tree representing an e-Service that allows for searching and buying mp3 files[5]. After an authentication step (action auth), in which the client provides *userID* and *password*, the e-Service asks for search parameters (e.g., author or group name, album or song title) and returns a list of matching files (action search); then, the client can: *(i)* select and listen to a song (interaction listen), and choose whether to perform another search or whether to add the selected file to the cart (action add_to_cart); *(ii)* add_to_cart a file without listening to it. Then, the client chooses whether to perform those actions again. Finally, by providing its payment method details the client buys and downloads the content of the cart (action buy).

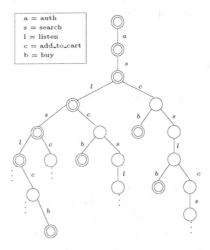

Fig. 1. Example of external execution tree of an e-Service

---

[4] Typically, in an e-Service, the root is final, to model that the computation of the e-Service may not be started at all by the client.

[5] Final nodes are represented by two concentric circles.

Note that, after the action **auth**, the client may quit the *e*-Service since he may have submitted wrong authentication parameters. On the contrary, the client is forced to buy, within the single interaction **buy**, a certain number of selected songs, contained in the cart, possibly after choosing and listening to some songs zero or more times.   □

3.2   Internal Schema

The internal schema maintains, besides the behavior of the *e*-Service, the information on which *e*-Services in the community execute each given action of the external schema. As before, here we abstract from the specific formalism chosen for giving such a specification, instead we concentrate on the notion of internal execution tree. Formally, each edge of an internal execution tree of an *e*-Service E is labeled by (a, I), where a is the executed action and I is a nonempty set denoting the *e*-Service instances executing a. Every element of I is a pair (E', e'), where E' is an *e*-Service and e' is the identifier of an instance of E'. The identifier e' uniquely identifies the instance of E' within the internal execution tree. In general, in the internal execution tree of an *e*-Service E, some actions may be executed also by the running instance of E itself. In this case we use the special instance identifier **this**. Note that the execution of each action can be delegated to more than one other *e*-Service instance.

An internal execution tree induces an external execution tree: given an internal execution tree t_i we call *offered external execution tree* the external execution tree t_e obtained from t_i by dropping the part of the labeling denoting the *e*-Service instances, and therefore keeping only the information on the actions. An internal execution tree t_i *conforms to* an external execution tree t_e if t_e is equal to the offered external execution tree of t_i. An *e*-Service is *well formed* if its internal execution tree conforms to its external execution tree.

We now formally define when an *e*-Service of a community correctly delegates actions to other *e*-Services of the community. We need a preliminary definition: given an internal execution tree t_i of an *e*-Service E, and a path p in t_i starting from the root, we call the *projection* of p on an instance e' of an *e*-Service E' the path obtained from p by removing each edge whose label (a, I) is such that I does not contain e', and collapsing start and end node of each removed edge.

We say that the internal execution tree t_i of an *e*-Service E is *coherent* with a community C if:

- for each edge labeled with (a, I), the action a is in the alphabet of C, and for each pair (E', e') in I, E' is a member of the community C;
- for each path p in t_i from the root of t_i to a node x, and for each pair (E', e') appearing in p, with e' different from **this**, the projection of p on e' is a path in the external execution tree t'_e of E' from the root of t'_e to a node y, and moreover, if x is final in t_i, then y is final in t'_e.

Observe that, if an *e*-Service of a community C is simple, i.e., it does not delegate actions to other *e*-Service instances, then it is trivially coherent with C. Otherwise, i.e., it is composite and hence delegates actions to other *e*-Service instances, the behavior of each one of such e-Service instances must be correct according to its external schema.

A community of e-Services is *well-formed* if each e-Service in the community is *well-formed*, and the internal execution tree of each e-Service in the community is coherent with the community.

Example 2. Figure 2 shows an internal execution tree, conforming to the external execution tree in Figure 1, where the `listen` action is delegated to a different e-Service, using each time a new instance. The internal execution tree, conforming again to the external execution tree in Figure 1, where no action is delegated to other e-Service instances, is characterized by the edges labeled by $(a, E, this)$, being α any action.

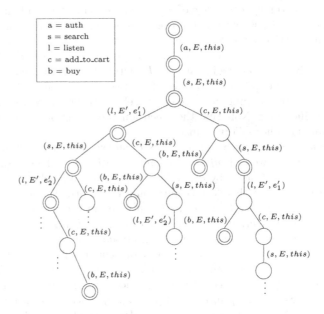

Fig. 2. Example of internal execution tree of a composite e-Service

In the examples each action is either executed by the running instance of E itself, or is delegated to exactly one other instance. Hence, for simplicity, in the figure we have denoted a label $(a, \{(E, e)\})$ simply by (a, E, e).   □

4 e-Service Instances

In order to be executed, a deployed e-Service has to be activated, i.e., necessary resources need to be allocated. An e-Service instance represents such an e-Service running and interacting with its client.

From an abstract point of view, a running instance corresponds to an execution tree with a highlighted node, representing the "current position", i.e., the point reached by the execution. The path from the root of the tree to the current position is the run of the e-Service so far, while the execution (sub-)tree having as root the current position describes the behavior of what remains of the e-Service once the current position is reached.

Formally, an e-Service instance is characterized by:

- an *instance identifier*,
- an *external view* of the instance, which is an external execution tree with a current position,
- an *internal view* of the instance, which is an internal execution tree with a current position.

Example 3. Figure 3 shows an external view of an instance of the e-Service of Figure 1. The sequence of actions executed so far and the current position on the execution tree are shown in thick lines. It represents a snapshot of an execution by a client that has provided its credentials and search parameters, has searched for and listened to one mp3 file, and has reached a point where it is necessary to choose whether *(i)* performing another search, *(ii)* adding the file to the cart, or *(iii)* terminating the e-Service (since the current position corresponds to a final node). □

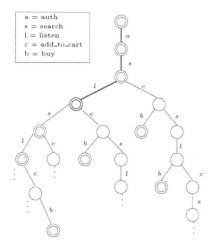

Fig. 3. External view of an e-Service instance

The internal view of an e-Service instance additionally maintains information on which e-Service instances execute which actions. At each point of the execution there may be several other active instances of e-Services that cooperate with the current one, each identified by its instance identifier. Note that, in general, an action can be executed by one or by more than one e-Service instance. The opportunity of allowing more than one component e-Service to execute the same action is important in specific situations, as the one reported in [4].

5   Running an *e*-Service Instance

In Section 2 we have briefly shown the steps that a client should perform in order to execute an *e*-Service, namely:

1. activation of the *e*-Service instance,
2. choice of the invokable actions
3. termination of the *e*-Service instance,

where step (*2*) can be performed zero or more times, and steps (*1*) and (*3*) only once. Each of these steps is constituted by sub-steps, consisting in executing commands and in sending acknowledgements, each of them being executed by a different actor (either the client or the *e*-Service).

In what follows we describe the correct sequence of interactions between a client and an *e*-Service, assuming, for the sake of simplicity, that no action is executed simultaneously by different *e*-Services (see Section 4). It is easy to extend what presented in order to cover also this case. Figure 4 shows the conceptual interaction protocol.

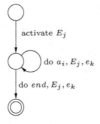

(a) Client

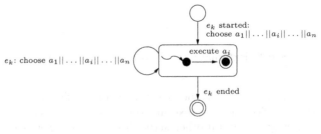

(b) *e*-Service

Fig. 4. Conceptual Interaction Protocol

Activation. This step is needed to create the *e*-Service instance. The client[6] invokes the activation command, specifying the *e*-Service to interact with. If E_j is such an *e*-Service, the syntax of this command is:

activate E_j

When this command is invoked, all the necessary resources for the execution of a new instance e_k of *e*-Service E_j are allocated. Additionally, each *e*-Service instance creates a copy of both the internal and the external execution tree characterizing the *e*-Service schema it belongs to.

As soon as e_k is ready to execute, it responds to the client with the message

e_k **started: choose** $a_1 || a_2 || \ldots || a_n$

The purpose of this message is threefold. First, the client has an acknowledgement that the invoked *e*-Service has been activated and that the interactions may correctly start. Second, the client is informed about the instance identifier he will interact with (e_k). Third, the client is asked to choose the action to execute among $a_1, \ldots, a_n$. The choice command is described next.

Choice. This step represents the interactions carried on between the client and the *e*-Service instance. Each *e*-Service instance is characterized, wrt the client, by its external execution tree, and all the actions are offered according to the information encoded in such a tree. Therefore, according to its external execution tree, the *e*-Service instance e_k proposes to its client a set of possible actions, e.g., $a_1, \ldots, a_n$, and asks the client to choose the action to execute next among $a_1, \ldots, a_n$. The syntax of this command is:

e_k**: choose** $a_1 || a_2 || \ldots || a_i || \ldots || a_n$

where $||$ is the choice symbol.

According to his goal, the client makes his choice by sending the message

do a_i, E_j, e_k

In this way, the client informs the instance e_k of *e*-Service E_j that he wants to execute next the action a_i. Once e_k has received this message, it executes action a_i. The execution of a_i is transparent to the client: the latter does not know anything about it, it only knows when it is ended, i.e., when the *e*-Service asks him to make another choice. This is shown in Figure 4 by the composite state that contains a state diagram modeling the execution of a_i.

The role of E_j and e_k becomes especially clear if we consider that the client could be a composite *e*-Service. When a composite *e*-Service E delegates an action to a component *e*-Service (e.g., E_j), it needs to activate a new *e*-Service instance (e_k), thus becoming in its turn a client. Therefore, on one side, E

---

[6] The client may be either a human user or another *e*-Service, however, for the sake of simplicity, in what follows we consider a human client.

interacts with the *external* instances of the component e-Service, since E is a client of the latter; on the other side, E chooses which action is to be invoked on which e-Service (either itself or a component e-Service) according to its internal execution tree, when E acts as "server" towards its client.

Termination. Among the set of invokable actions there is a particular action, **end**, which, if chosen, allows for terminating the interactions. Therefore, if the current node on the external execution tree is a final node, the e-Service proposes a choice as:

$$e_k\text{: \textbf{choose} } end||a_1||a_2||\ldots||a_i||\ldots||a_n$$

and if the client has reached his goal, he sends the message:

$$\textbf{do } end, E_j, e_k$$

The purpose of this action it to de-allocate all the resources associated with instance e_k of e-Service E_j. As soon as this is done, the e-Service informs its client of it with the message:

$$e_k\text{: \textbf{ended}}$$

Examples of interactions can be found in [3].

6 Composition Synthesis

When a user requests a certain service from an e-Service community, there may be no e-Service in the community that can deliver it directly. However, it may still be possible to synthesize a new composite e-Service, which suitably delegates action execution to the e-Services of the community, and when suitably orchestrated, provides the user with the service he requested. Hence, a basic problem that needs to be addressed is that of e-Service *composition synthesis*, which can be formally described as follows: given an e-Service community C and the external execution tree t_e of a target e-Service E expressed in terms of the alphabet of C, synthesize an internal execution tree t_i such that *(i)* t_i conforms to t_e, *(ii)* t_i delegates all actions to the e-Services of C (i.e., **this** does not appear in t_i), and *(iii)* t_i is coherent with C.

Figure 5 shows the architecture of an e-*Service Integration System* which delivers possibly composite e-Services on the basis of user requests, exploiting the available e-Services of a community C. When a client requests a new e-Service E_0, he presents his request in form of an external e-Service schema $t_e^{E_0}$ for E_0, and expects the e-Service Integration System to execute an instance of E_0. To do so, first the *composer* module makes the composite e-Service E_0 available for execution, by synthesizing an internal schema $t_i^{E_0}$ of E_0 that conforms to the external schema $t_e^{E_0}$ and is coherent with the community C. Then, using the internal schema $t_i^{E_0}$ as a specification, the *orchestration engine* activates an (internal) instance of E_0, and orchestrates the different available e-Services, by activating

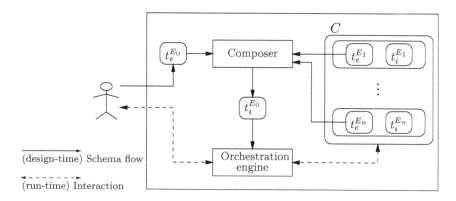

Fig. 5. e-Service Integration System

and interacting with their external view, so as to fulfill the client's needs. The orchestration engine is also in charge of terminating the execution of component e-Service instances, offering the correct set of actions to the client, as defined by the external execution tree, and invoking the action chosen by the client on the e-Service that offers it.

All this happens in a transparent manner for the client, who interacts only with the e-Service Integration System and is not aware that a composite e-Service is being executed instead of a simple one.

7   Conclusions

In this paper we have proposed a conceptual, and formal, vision of e-Services, in which we distinguish between the external behavior of an e-Service as seen by clients, and the internal behavior as seen by a deployed application running the e-Service, which includes information on delegation of actions to other e-Services. Such a vision clarifies the notion of composition from a formal point of view. On the basis of such a framework, we will study techniques for automatic composition synthesis.

Note that in the proposed framework, we have made the fundamental assumption that one has complete knowledge on the e-Services belonging to a community, in the form of their external and internal schema. We also assumed that a client gives a very precise specification (i.e., the external schema) of an e-Service he wants to have realized by a community. In particular, such a specification does not contain forms of "don't care" nondeterminism. Both such assumptions can be relaxed, and this leads to a development of the proposed framework that is left for further research.

Acknowledgments

This work has been partially supported by MIUR through the "Fondo Strategico 2000" Project *VISPO* and the "FIRB 2001" Project *MAIS*. The work of Massimo

Mecella has been also partially supported by the European Commission under Contract No. IST-2001-35217, Project *EU-PUBLI.com*.

References

1. M. Aiello, M.P. Papazoglou, J. Yang, M. Carman, M. Pistore, L. Serafini, and P. Traverso, *A Request Language for Web-Services Based on Planning and Constraint Satisfaction*, Proceedings of the 3rd VLDB International Workshop on Technologies for e-Services (VLDB-TES 2002), Hong Kong, China, 2002.
2. A. Ankolekar, M. Burstein, J. Hobbs, O. Lassila, D. Martin, D. McDermott, S. McIlraith, S. Narayanan, M. Paolucci, T. Payne, and K. Sycara, *DAML-S: Web Service Description for the Semantic Web*, Proceedings of the 1st International Semantic Web Conference (ISWC 2002), Chia, Sardegna, Italy, 2002.
3. D. Berardi, D. Calvanese, G De Giacomo, M. Lenzerini, and M. Mecella, *A Foundamental Framework for e-Services*, Technical Report 10-03, Dipartimento di Informatica e Sistemistica, Università di Roma "La Sapienza", Roma, Italy, 2003, (available on line at: http://www.dis.uniroma1.it/~berardi/publications/techRep/TR-10-2003.ps.gz).
4. D. Berardi, D. Calvanese, G De Giacomo, and M. Mecella, *Composing e-Services by Reasoning about Actions*, Proc. of the ICAPS 2003 Workshop on Planning for Web Services, 2003.
5. D. Berardi, L. De Rosa, F. De Santis, and M. Mecella, *Finite State Automata as Conceptual Model for e-Services*, Proc. of the IDPT 2003 Conference, 2003, To appear.
6. F. Casati and M.C. Shan, *Dynamic and Adaptive Composition of e-Services*, Information Systems **6** (2001), no. 3.
7. D. Georgakopoulos (ed.), *Proceedings of the 9th International Workshop on Research Issues on Data Engineering: Information Technology for Virtual Enterprises (RIDE-VE'99)*, Sydney, Australia, 1999.
8. R. Hull, M. Benedikt, V. Christophides, and J. Su, *E-Services: A Look Behind the Curtain*, Proceedings of the 22nd ACM SIGACT-SIGMOND-SIGART Symposium on Principles of Database Systems (PODS), June 2003.
9. S. McIlraith and T. Son, *Adapting Golog for Composition of Semantic Web Services*, Proceedings of the 8th International Conference on Knowledge Representation and Reasoning (KR 2002), Toulouse, France, 2002.
10. M. Mecella and B. Pernici, *Building Flexible and Cooperative Applications Based on e-Services*, Technical Report 21-2002, Dipartimento di Informatica e Sistemistica, Università di Roma "La Sapienza", Roma, Italy, 2002, (available on line at: http://www.dis.uniroma1.it/~mecella/publications/mp_techreport_212002.pdf).
11. B. Medjahed, B. Benatallah, A. Bouguettaya, A.H.H. Ngu, and A.K. Elmagarmid, *Business-to-Business Interactions: Issues and Enabling Technologies*, VLDB Journal **12** (2003), no. 1.
12. J. Yang and M.P. Papazoglou, *Web Components: A Substrate for Web Service Reuse and Composition*, Proceedings of the 14th International Conference on Advanced Information Systems Engineering (CAiSE'02), Toronto, Canada, 2002.
13. J. Yang, W.J. van den Heuvel, and M.P. Papazoglou, *Tackling the Challenges of Service Composition in e-Marketplaces*, Proceedings of the 12th International Workshop on Research Issues on Data Engineering: Engineering E-Commerce/E-Business Systems (RIDE-2EC 2002), San Jose, CA, USA, 2002.

Querying Spatial Resources. An Approach to the Semantic Geospatial Web

J. E. Córcoles and P. González

Departamento de Informática, Escuela Politécnica Superior de Albacete.
Universidad de Castilla-La Mancha
02071, Albacete (Spain)
{corcoles, pgonzalez}@info-ab.uclm.es

Abstract. The research community has begun an effort to investigate foundations for the next stage of the Web, called *Semantic Web* [1]. Current efforts
include the Extensible Markup Language XML, the Resource description
Framework, Topic Maps and the DARPA Agent Markup Language
DAML+OIL. A rich domain that requires special attention is the semantics of
geospatial information. In order to approach the Semantic Geospatial Web, a
mediation system for querying spatial and non-spatial information is presented
in this paper. The spatial information is represented in the Geographical
Markup Language (GML). GML is an XML encoding developed by the Open-
GIS Consortium for the transport and storage of spatial/geographic information,
including both spatial features and non-spatial features. In order to integrate
spatial information from several spatial XML documents (GML), we have
based our work on the *Community Web Portal* concept with RDF and RQL, a
declarative language to query both RDF descriptions and related schemas.

1 Introduction

With the growth of the World Wide Web has come the insight that currently available
methods for finding and using information on the Web are often insufficient. Today's
retrieval methods are typically limited to keyword searches or matches of sub-string,
offering no support for deeper structures that might lie hidden in the data or that people typically use to reason; therefore, users may often miss critical information when
searching the Web. At the same time, the structure of the posted data is flat, which in-
creases the difficulty of interpreting the data consistently. There would exist a much
higher potential for exploiting the Web if tools were available that better match human reasoning. In this vein, the research community has begun an effort to investigate
foundations for the next stage of the Web, called *Semantic Web* [1]. Current efforts
include the Extensible Markup Language XML, the Resource description Framework,
Topic Maps and the DARPA Agent Markup Language DAML+OIL.

A rich domain that requires special attention is the Semantics of Geospatial Information [2]. The enormous variety of encoding of geospatial semantics makes it
particularly challenging to process requests for geospatial information. Work led by
the OpenGIS Consortium [3] addressed some basic issues, primarily related to the
geometry of geospatial features. The Geography Markup Language (GML) provides a

C. Bussler et al. (Eds.): WES 2003, LNCS 3095, pp. 41–50, 2004.

syntactic approach to encoding geospatial information through a language in which symbols need to be interpreted by users, because associated behaviour is not accounted for.

In order to approach the Semantic Geospatial Web, a mediation system for querying spatial and non-spatial information is presented in this paper. The spatial information is represented in the Geographical Markup Language (GML). GML is an XML encoding for the transport and storage of spatial/geographic information, including both spatial features and non-spatial features [4]. In this way, the proposed architecture integrates different approaches for querying GML on the Web.

Query mediation has been extensively studied in the literature for different kinds of mediation models and for the capabilities of various sources. In the ambit of non-spatial integration there are several approaches such as *Tsimmis* [5], *YAT*[6], *Information Manifold* [7], *PICSEL* [8] and *C-Web* [9]. More specifically, in spatial data integration, there are approaches developed by [10] and [11]. [10] extends the MIX wrapper-mediator architecture to perform spatial data integration. [11] presents a mediation system that addresses the integration of GIS data tools, following a GAV(Global as View) approach. On the other hand, there are projects related to Geospatial Portals: *Geospatial One Stop Portal* [12] and *G-Portal* [13].

We have based our work on the C-Web Portal [9] for supporting the integration of spatial (expressed in GML documents) and non-spatial information. C-Web, which provides the infrastructure for (1) publishing information sources and (2) formulating structured queries by taking into consideration the conceptual representation of a specific domain in the form of an ontology, follows a LAV (Local As View) approach. The features of C-Web (RDF, RQL, ...) applied to search for spatial information the main contribution of this work, and it offers a closer focus on the Semantic GeoSpatial Web.

This paper is organized as follows: Section 2 describes the main aim of this work. Section 3 gives an overview of the proposed architecture. The conclusions and future work are presented in Section 4.

2   Domain of the Problem

A *Community Web Portal* essentially provides the means to select, classify and access, in a semantically meaningful and ubiquitous way, various information resources (sites, documents, GML documents, data) for diverse target audiences (corporate, inter-enterprise, ...). The core Portal component is a *Catalog* holding descriptions, i.e. metadata, about the resources available to the community members. In order to effectively disseminate community knowledge, Portal Catalogs organize and gather information in a *multitude of ways*, which are far more flexible and complex than those provided by standard (relational or object) databases. We used the Resource Description Framework (RDF) standard [14] [15] proposed by W3C, designed to facilitate the creation and exchange of resource descriptions between Community Webs. In order to query the *Catalog*, a query language, called RQL, is presented in [16] which allows semistructured RDF descriptions to be queried using taxonomies of node and edge labels defined in the RDF schema.

In order to integrate the spatial information of several spatial XML documents (GML), we have based our work on the *Community Web Portal* concept with RDF and RQL, a declarative language to query both RDF descriptions and related schemas

[9]. To perform this integration, it is necessary to make some modifications to the original approach because in the original approach the *Catalog* is considered as a collection of resources identified by URIs and it is described using properties. However, it does not need to use operator over the resources, only over the properties.

GML documents (or part of) are a resource. Unlike the original approach, it is possible to apply spatial operators (comparatives: cross, overlap, touch; analysis: Area, Length) over the resources provided they represent geometry information with GML. In order to take advantage of this fact, we have designed two modifications with respect to the original approach:

✐✂Extension of RQL to support spatial operators over the resources that represent spatial documents or part of spatial documents. These operators must be the same as those defined in [17] for our query language over GML. There are two types of operators: methods for testing Spatial Relations and methods that support Spatial Analysis. This extension is not dealt with in this paper. We assumed that this extension had been carried out.

✐✂Extension of the *Community Web Portal* architecture to support the application of the spatial operators over the resources involved in the query, and the integration of all information to be returned. The definition of these new components in the Catalog architecture is the main aim of this paper.

2.1 An Overview

In this section, an overview of the application of our mediation system is given, looking at the system from the point of view of the user. In Figure 1, an example of the Portal Schema and its instances is shown. The example has been obtained from

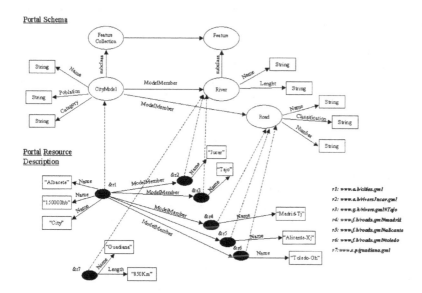

Fig. 1. Portion of Catalog of a CityModel

the specification documents of GML by OpenGIS [18]. Due to RDF's capability for adding new feature and geometry types in a clear and formal manner [18], this example has been carried out extending the geospatial ontology defined by OpenGIS, where the class (Geometry, LineString, etc) and properties (coordinates, Polygon-Member, etc) are defined. The example shows an extension of a Catalog for a City-Model proposed by [18] and called Cambridge.rdfs. This is a well-known example used in all specifications over GML developed by OpenGIS. The Cambridge example has a single feature collection of type 'CityModel' and contains two features using a containment relationship called 'modelMember'. The model member can be Rivers that run through the City or Roads belonging to the City.

An example of a query expressed in RQL extended with spatial operators may be as follows:

> *"Find all Road resources that belong to Albacete City and which are within 50 meters of a River of this city ".*

In RQL:

> ***Select** Z*
> ***From** {X}ModelMember{Y:River}, {W}ModelMember{Z:Road}, {P}name{Q}*
> ***Where** X=W and X=P and Q="Albacete" and Crosses(Buffer(Y,50),Z)*

In a different way to the original approach, this query has a part that is executed directly over the *Catalog*, and another part that is executed over the objects *Road* and *River* stored in the respective sources. In order to do this, it is necessary to establish a spatial query plan. On the other hand, if the query uses operators like *Area, Length, Union, Intersection*, etc., this approach manages the new created resources. The results could be the resources (&r2, &r3).

Although the resources may be of different types (documents, HTML files, Raster image,..) in this approach the semantic of the spatial operators is only applied over the geometry objects based on the OpenGIS specification [3].

2.2   Query Language over GML Document

This Section describes the main features of our query language over GML documents. It was developed as a spatial query language over GML documents. The data model and the algebra underlying the query language are defined in [19]. The query language has a familiar *select-from-where* syntax and is based on SQL. It includes a set of spatial operators (disjoint, touches, etc.), and includes traditional operators (=,>, <,...) for non-spatial information. The characteristics of our language are:

Path Expressions: The dot notation is used to navigate in the data model.

Ability to return an XML document: For compatibility with other tools, the result set of a query is another XML document. The structure of the new document should respect the GML schemas if it has geometry types.

Ability to query and return XML tags and attributes: This language allows some attributes and tags to be partially defined in the query. The symbol '%' is used as a wildcard in the partial definitions.

Intelligence type coercion: This language supports automatic coercion between simple types. However, coercion is not defined for geometry types.

Handles unexpected data: To define this feature, the operators 'Like' and '=' are handled in the same way as an existential quantifier. Thus, if a tag has several values, it then allows the first to be selected to carry out the condition.

Ability to allow queries when the schema is not fully known: In order to make it unnecessary for the user to know the schema of a document, this language shows the optional tags and attributes in brackets, each one of them being separated by the symbol 'I'. This symbol 'I' means "optimality between different tags or attributes. In this way, the user indicates that he or she wants to find values stored in elements with a different syntax but with the same semantics.

Returns unnamed attributes: The language allows us to establish in the *Select* clause whether it returns only tags or tags with all these children. To achieve this the symbol '^' is used. This symbol means that a given tag or attribute is not included in the result set.

Allows Geometry operators: This feature is the main difference with respect to other query languages. There are two types of operators: methods for testing Spatial Relations and methods that support Spatial Analysis.

3 Architecture

In [16] the existence of the following countably infinite and pairwise disjoint sets of symbols is assumed: C of Class name, P of Property names, U of Resource URIs as well as a set L of Literal type names like *string, integer, date*, etc. Focusing on the Resources, the RQL queries obtain all resources that satisfy certain conditions. In this case, all data necessary for checking the query are available in the Catalog (Portal Schema and Portal Resource Description). However, in order to add spatial features to the Catalog, it is necessary to take these premises into account:
    The spatial features that describe a resource must not be included in the Portal Resource Description, like the non-spatial features, because this means very large Catalogs and very difficult for managing. For this reason, in our approach, the spatial features are kept (preserved) at the local source.
    In this way, we have included new components for allowing this kind of queries where the non-spatial features are executed over the Catalog, and the spatial part is executed over one or several spatial resources.
    The spatial resources are represented by GML because it is an XML encoding for the transport and storage of spatial/geographic information, including both spatial features and non-spatial features. The mechanisms and syntax that GML uses to encode spatial information in XML are defined in the specification of OpenGIS [20]. OGC

manages consensus processes that result in interoperability between diverse geoprocessing systems. Therefore, GML proposes a standard format to represent the spatial information with XML. The main aim of using GML in this approach is that it can be queried using a specific query language [17] and it stores efficiently [21]. These are two important features of application in geographic information systems.

In addition, although the GML document follows the Spatial specification developed by OpenGIS, the models used for each source for representing the spatial information can be different. For example, source S1 and S2 can represent the same information about Roads, but S1 can represent Road as a single LineString and S2 can represent Road as a set of several contiguous lane(s) going in the same direction. For this reason, a mapping between XML fragments and ontology concepts and roles is necessary. We assumed that the mapping is not necessary over the non-spatial features.

The policy of returning the results of a query are also different. When the query carries out an integration of information from several sources or Spatial Analysis operators, the resource returns are *virtual resources* (temporal or cached) because the original resource has been updated by these operators.

In Figure 2, the architecture of our approach is shown (the component related to C-Web, except *Catalog*, has been omitted). Basically, it is composed of a query processor that consists of three components: *Analyzer*, *Optimizer* and *Execution module*. There is also a wrapper with the task of executing the partial queries in each source.

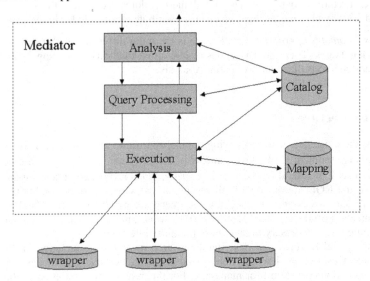

Fig. 2. Architecture

Analyzer: The *Analyzer* component consists of (1) the parser, analysing the syntax of queries; (2) the graph constructor, capturing the semantic of queries in terms of typing [16] and interdependencies of expressions involved; and (3) the evaluation engine of RQL, accessing the RDF description from the underlying database (note that in RQL implementation, RDF is stored over an object relational schema). In [16] it is the main component for querying with RQL, but in our approach it is necessary to add new components to try to query the spatial information. With this architecture it is only possible to query with alphanumeric operators. The result of the *Analyzer* component is a query expressed in a query language for an ORDBMS (SQL). This is the starting point in the execution of queries with spatial operators. The spatial operators expressed in RQL are not translated to the syntax and semantic of SQL, because this query will not be executed directly over the ORDBMS when it has spatial operators.

Query Processing: This query is the input of the *Query Processing* component. There is a query in this component because it has spatial operators. In this component the query must be treated in a different way for spatial information and for non-spatial information, due to the fact that all non-spatial information is stored in a DBMS (Catalog) and the spatial information is stored in different GML documents in different locations. However, the Catalog provides the relation between the non-spatial data and the spatial data. This model is comparable with approaches of a spatial database system in the manner in which spatial and non-spatial data are stored and linked to each other, for example SAND [22]. SAND (Spatial And Non-spatial Data) assumes the spatial description of objects is stored in disk-based spatial data structures which are linked properly to the rest of the objects' non-spatial description. In order to process spatial queries from this point of view, SAND presents a variety of feasible strategies for answering spatial and mixed queries. Due to the similarity between both schemas, the optimisation strategies offered by SAND can be applied to this approach. These strategies cover everything from *un-optimized* strategies to *further pipelining* strategies [22].

Execution: The query plan is executed in the *Execution* component. The non-spatial attributes are obtained from the Catalog, whereas the spatial operators are executed over each spatial resource involved. The component has two tasks:

(1) The user views the Portal as a simple database of fragment without knowledge of the source on which each fragment is located. We might then consider each fragment as an object whose identity is the location path of the fragment. A mapping allows the organization of the fragments into a collection of instances of concepts in the Ontology. It follows that one can in principle query this database of fragments using a query language. The answer of the query is defined using the semantic of queries on object bases. However, even though the answer is well-defined, an efficient evaluation requires that we use the mapping effectively to translate the query into one or more queries on the sources. In order to do this, we have based our work on our previous work [23]. In this paper, a language for stabilizing mappings between XML fragments and ontology concepts and roles is presented. In [23] we have used ontologies to solve the schematic heterogeneity between different GML documents. We use RDF to know how a schema satisfies an ontology. These schemas are always expressed as DTDs for each site.

We have used our query language over GML documents to be executed in the wrappers. It is a query language over GML/XML enriched with spatial operators [19]. This query language has an underlying data model and algebra that supplies the semantics of the query language [17].

(2) The second task is to integrate the results of the queries executed in each source [24]. As is mentioned above, when Spatial Analysis operators (like *Area*, *Length*, etc) are used in the query, or operators like *Union*, *Intersect*, *Difference*, etc., it is necessary to generate in this component a GML document with this new data. In this way, the resulting resources of a query can be *fictitious* resources because they can be stored (cached) in the Portal.

Wrapper: The most important task of this component is to execute the query over the GML document that contains the objects searched. The query must be adapted to the underlying data model for each document [17]. In order to execute our query language over GML documents, an efficient storage of GML documents is necessary. In [21] we tend to use approaches based on relational databases to store GML documents. In this way, we can use a complete set of data management services (including concurrency control, crash recovery, scalability, etc) and benefit from the highly optimized relational query processor. In addition, RDBMS allows us to store spatial objects in accordance with the possibilities offered in [25].

4 Conclusions

In this paper a mediation system for querying spatial and non-spatial information is presented. We have based this work on the C-Web Portal [9] for supporting the integration of spatial (expressed in GML documents) and non-spatial information. The main feature of this approach is the possibility of applying spatial operators over the resources that are represented by GML format. Thus, with this approach, it is possible to discover spatial and non-spatial resources interrelated semantically on the Web. In this way, this work contributes a small step towards the Semantic Geospatial Web.

Future work foresees the extension of the *Loader* developed by c-Web for supporting the load of spatial resources. It is a most important point for obtaining an efficient and operative system with these features.

References

1. J. Berners-Lee, J. Hendler and O. Lassila. The Semantic Web, Scientific American, vol 184, no. 5, pp. 34–43, 2001.
2. J. Egenhofer. Toward the Semantic Geospatial Web. ACM-GIS 2002. 10th ACM International Symposium on Advances in Geographic Information Systems. McLean (USA). 2002.
3. OpenGis Consortium. Specifications. http://www.opengis.org/techno/specs.htm.1999
4. OpenGIS. Geography Markup Language (GML) v3.0. http://www.opengis.org/techno/documents/02-023r4.pdf. 2003.
5. Y. Papakonstantinou, H. García-Molina and J. Widom. Object Exchange Across Heterogeneous Information Sources. In Proc. ICDE Conf. TSIMMIS project: http://www-db.standford,edu/tsimmis. 1995.
6. D. Christophides, S. Cluet and J. Siméon. On Wrapping Query Languages and Efficient XML integration. In Proc. Of ACM SIGMOD Int. Conference on Management of Data Dallas (USA). 2000.

7. A. Y. Levy, A. Rajaraman and J. Ordille. Querying Heterogeneous Information Sources Using Source Description. In Proc. of the Int. Conference on Very Large Databases, pp. 25–262. India. 1996.

8. F. Goasdoué, V. Lattés and M-C Rousset. The use of CARIN Language and Algorithms for Infomration Integration: The PICSEL System. International Journal on Cooperative Information Systems. 2000.

9. B. Amann, I. Fundulaki and M. Scholl. Integrating Ontologies and thesauri for RDF schema creation and metadata querying. International Journal of Digital Libraries. 2000.

10. A. Gupta, R. Marciano, I. Zaslavsky and C. Baru. Integrating GIS and Imagenery through XML based information Mediation. Integrated Spatial Databases: Digital Images and GIS. Lecture Notes in Computer Science. Vol. 1737, pp. 211–234. Springer-Verlag. 1999.

11. O. Boucelma, M. Essid and Z. Lacroix. A WFS-Based Mediation System for GIS Interoperability. ACM-GIS 2002. 10th ACM International Symposium on Advances in Geographic Information Systems. McLean (USA). 2002.

12. GEOSpatial One-Stop Portal. http://ip.opengis.org/gos-pi/. 2002.

13. E.P. Lim, D. Goh, Z. Liu, W.K. Ng, C. Khoo, S.E. Higgins, G-Portal: A Map-based Digital Library for Distributed Geospatial and Georeferenced Resources, Proc. of the Second ACM+IEEE Joint Conference on Digital Libraries (JCDL 2002), USA. 2002.

14. D. Brickley and R.V. Guha. Resource Description Framework (RDF) Schema Specification 1.0, W3C Candidate Recommendation. Technical Report CR-rdf-schema-20000327, W3C, 2000. Available at http://www.w3.org/TR/rdf-schema.

15. O. Lassila and R. Swick. Resource Description Framework (RDF) Model and Syntax Specification. W3C Recommendation, February 1999. Available at http://www.w3.org/TR/REC-rdf-syntax.

16. S. Alexaki, V. Christophides, G. Karvounarakis, D. Plexousakis, K. Tolle. "*The ICSFORTH RDF Suite: Managing Voluminous RDF Description Bases*". In Proceedings of the 2nd International Workshop on the Semantic Web (SemWeb'01), in conjunction with WWW10, pp. 1–13, Hong Kong. 2001.

17. J. Córcoles and P. González. A Specification of a Spatial Query Language over GML. ACM-GIS 2001. 9th ACM International Symposium on Advances in Geographic Information Systems. Atlanta (USA). 2001.

18. OpenGIS. Geography Markup Language (GML) v3.0. http://www.opengis.org/techno/documents/02-023r4.pdf. 2003.

19. J. Córcoles and P. González. A spatial query language over XML documents. Fifth IASTED International Conference on Software Engineering and Applications (SEA). pp. 1–6. 2001.

20. OpenGis Consortium. Specifications. http://www.opengis.org/techno/specs.htm. 1999.

21. J. Córcoles and P. González. Analysis of Different Approaches for Storing GML Documents ACM-GIS 2002. 10th ACM International Symposium on Advances in Geographic Information Systems. McLean (USA). 2002.

22. W. G. Aref and H. Sament. Optimization Strategies for Spatial Query Processing. Proc. 17th International Conference on Very Large Data Bases. Barcelona, 1991.

23. J. Córcoles and P. González. A First Approach for Querying Spatial XML documents with RDF. Accept in the International Conference on Web Engineering. Oviedo (Spain). July 2003. To Appear in Lecture Notes in Computer Science (LNCS) *by Springer-Verlag*. 2003.

24. A. Gupta, I. Zaslavsky and R. Marciano. Generating query evaluations plans within a spatial mediation framework. Proceedings of the 9th International Symposium on Spatial Data Handling. 2000.
25. Open GIS Consortium, Inc. OpenGIS: Simple Features Specification For SQL Revision 1.1 OpenGIS 99-049 Release. 1999.

A Framework for E-markets: Monitoring Contract Fulfillment

Lai Xu

Tilburg University, CRISM/Infolab,
5000 LE, Tilburg, The Netherlands
L.Xu@uvt.nl
http://infolab.uvt.nl/people/lxu

Abstract. To be able to monitor contracts at contract fulfillment stages is a key ingredient for a reliable, flexible, efficient, realistic and acceptable e-market. In this paper, we explore monitoring requirements of different e-market infrastructures, and define the monitorability of various e-market infrastructure approaches. For avoidance and anticipation of imminent contract violations, and detection and compensation of actual violations, a two-level framework is presented, a reference architecture of the two-level framework is demonstrated finally.

1 Introduction

To be able to monitor multi-partner contracts at the contract fulfillment stage is the key for realizing a reliable, flexible, efficient, realistic and acceptable e-market, which benefits all the business parties involved. An agent-based e-commerce infrastructure is regarded as one of the most suitable open environments for electronic marketplaces [6]. Because of their autonomous, reactive and proactive features, agents can act on behalf of their owner and use individual strategies to increase the ability to do business; unfortunately, the problem of how to force agents to comply with prescribed behavior and thus achieve effective monitoring is complicated. Most existing e-commerce platforms implement logging, auditing and failure recovery functions [1] [5], but still miss avoidance and anticipation of anomalous actions, failure detections and issue compensations. This clearly limits the e-business applications to a small number of simple strict application scenarios.

In contract-driven business process automation, a *contract* records the agreed upon obligations of contracting parties in terms of business process conditions [9]. Generally a contract has two stages, the first stage is a *contract establishment stage*, which includes contract conception, preparation and negotiation activities. In this stage, the parties' roles, responsibilities, obligations and deliverabilities are identified [4]. The second stage is a *contract fulfillment stage*, which is related to the behavior of contractual parties and may include monitoring, enforcement and compensation activities [21][24] [20] [13]. Our main concern within this stage is the realization of pro-active monitoring contracts.

C. Bussler et al. (Eds.): WES 2003, LNCS 3095, pp. 51–61, 2004.
© Springer-Verlag Berlin Heidelberg 2004

For realizing pro-active monitoring, we investigated monitoring requirements for different infrastructure e-markets at the contract fulfillment stage. We explored the *monitorability* of various e-market infrastructure approaches. *Monitorability* refers to whether sufficient information and monitoring points are provided, so that the e-market can be effectively monitored. Based on the distinguished monitoring requirements and the analysis of different e-market infrastructure approaches, a two-level monitoring framework is proposed.

In this paper, we used a standard case study from the CrossFlow project (see Section 2 for more details). Section 3 describes the monitoring requirements for e-markets. Section 4 presents our two-level monitoring framework and explains the purpose of each level. A reference architecture is elaborated in Section 5. Finally, we present our conclusions of concerning monitoring in e-markets.

2   Case study

In explaining existing issues in the multi-partner contract monitoring, we have chosen a standard multi-partner scenario [18] that outlines the manner in which a car damage claim is handled by an insurance company (AGFIL). The contractual parties work together to provide a service level which facilitates efficient claim settlement. The parties involved are called Europ Assist, Lee Consulting Services, Garages and Assessors. Europ Assist offers a 24-hour emergency call answering service to policyholders. Lee C.S. coordinates and manages the operation of the emergency service on a day-to-day level on behalf of AGFIL. Garages are responsible for car repairs. Assessors conduct the physical inspections of damaged vehicles and agree on repair costs with the garages involved.

In Figure 1, the general process of a car insurance case is described as followed: the policyholder phones Euro Assist using a tollfree number to notify of a new claim. Euro Assist registers the information, suggests an appropriate garage, and notifies AGFIL which checks whether the policy is valid and covers this claim. After AGFIL receives this claim, AGFIL sends the claim details to

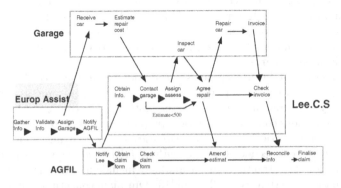

Fig. 1. The process diagram of car insurance [19]

Lee C.S. AGFIL subsequently sends a letter to the policyholder for a completed claim form. Lee C.S. agrees upon repair costs if an assessor is not required for small damages, otherwise an assessor is assigned. The assessor checks the damaged vehicle and agree upon repair costs with the garage. After receiving an repair agreement from Lee C.S., the garage then commences to repair. After finishing repairs, the garage issues an invoice to Lee C.S., which will check the invoice against the original estimate. Lee C.S. returns all invoices to AGFIL. AGFIL processes the payment. In the whole process, if the claim is found invalid, all contractual parties will be contacted and the process will be stopped. As we can be seen from this description, the workflow is complex between multiple partners and a multi-partner contract is used, this case could help to understand the issues of multi-partner contracts.

3   Monitoring Requirements for a Monitoring Framework

In Figure 2, we present a *Monitoring Contract Fulfillment Life Cycle* which can be split into two parts divided by the occurrence of an anomalous action [14] [16]. The part preceding the occurrence of anomalous actions is called the *proactive monitoring perspective*. The part following it is called the *reactive monitoring perspective*. In the pro-active monitoring stage, anomalous actions can be avoided and anticipated before contract violations occurrence. In the reactive monitoring stage, anomalous actions can be detected, the relevant partner needs to satisfactories compensated, and unsolvable disputations can be stored for future, human-involved resolution after the occurrence of contract violation.

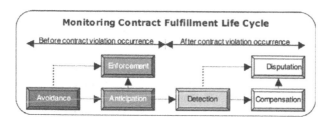

Fig. 2. The monitoring life cycle

After having described the monitoring contract fulfillment life cycle, we observe monitoring requirements from two perspectives. From **a pro-active monitoring perspective**, there are three monitoring requirements from the pro-active monitoring perspective:

– First, contractual parties need to be monitored for the purpose of avoidance and anticipation of contract violations. The execution of non-performance action needs to be forced. For example, in the car insurance case, after Europ

Assist assigned a garage for a policyholder who claimed a car damage before, Europ Assist also notified Lee C.S. and AGFIL. When Lee C.S. and AGFIL worked on this claim, Lee C.S. contacted the assigned garage and found that the policyholder did not send the car to the garage. The action *sending car* should be forced to perform in the car insurance process.

- Second, relevant events need to be recorded in a business process. After conflicts between contractual parties, these records can be used as evidence for what actually happened, who is/are responsible etcetera.
- Third, the execution of the actions needs to be measured to assure the quality of their performance. The measure depends on what actually partners did.

From **a reactive monitoring perspective**, the monitoring requirements from the reactive monitoring perspective are as followed.

- First, anomalous actions need to be detected after a contract violation. Particularly in a multi-party contractual business process an anomalous action is sometimes not detected until after other parties have performed many actions. Retrieval of certain activities of different parties is necessary. For example, in the car insurance case, the assigned garage did repair the car within the agreed amount with Lee C.S. and sent an invoice to Lee C.S., but the garage did not receive the repair cost from AGFIL. It can be caused by Lee C.S., because Lee did not forward the invoice, or AGFIL did not send a claim form to the policyholder, or the policyholder did not return the claim form to AGFIL etc. To detect who is/are responsible for the contract violation is most important in the reactive monitoring stage.
- Second, the costs of non-conforming actions or anomalous actions need to be compensated. Sometimes the compensation function is optional, but at a minimum the other parties need to be informed of the detection of anomalous actions to prevent further cost.
- Third, unsolvable disputations need to be stored for future human-involved arbitration and resolution.

A whole contract monitoring process should finish following function: anomalous action avoidance, imminent contract violation anticipation, non-conforming action enforcement, contract violation compensation and recording disputation. To satisfy the monitoring requirements we propose a two-level monitoring framework and explain how it can realize the monitoring requirements in next sections.

4   A Two-Level Monitoring Framework

We startly presenting a two-level framework. We explain why two levels of monitoring process specification are indeed required. In our view, there is a need for a framework with both a central monitoring level and a local monitoring level, as is shown in Figure 3. The *central monitoring level performs* overall monitoring on behalf of all contractual parties, whereas the *local monitoring level* only acts on behalf of its own party. Each party may adjust its monitoring request to accommodate different business processes. We have to emphasize that the central monitoring level and the local monitoring level are roles. It is not necessary to have a special agent/service to be the central partner. The central monitoring role could just as well be performed by a contractual partner who has an overall right to monitor the other contractual partners. For example, in our car insurance case, AGFIL could act as a central monitoring role.

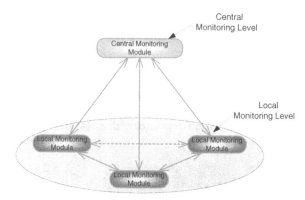

Fig. 3. A two-level framework

4.1   The Necessity of a Two-Level Monitoring

There exist different infrastructures for e-markets. E-market infrastructures can roughly be classified as: central control e-markets (e.g. contractual agent societies) and self-regulated e-markets [7] [23] [2]. At a centralized e-market, a central monitoring system can be employed, while at a decentralized e-market that is not possible. A centralized module can be embedded in a centralized control party; A local monitoring module is feasible for both e-market systems, it can be embedded in every peer party. All peer parties have their local monitoring module to automatically update their monitoring information.

While looking at an agent-mediated e-market, there are also different kinds of agents are that are distinguishable. Some agents are more independent than others; some agents use a remedial mechanism, which might return business processes to a normal course after occurrence of anomalous behavior. For example, when an agent changes business priorities of different business processes, it first deviates from the prescribed behavior for execution of another higher priority business process. Later it compensates this action and returns to normal behavior for the first business process. It is feasible that the agent might not want this kind of behavior to be monitored by a central monitoring agent and receive possible punishment.

In addition, agents within a society e-market may cooperate with outside agents to finish a business process. It is impossible for a society central monitoring agent to monitor those outside agents, but it is necessary that the business process is monitored by the local monitoring level. The question of how to keep a balance between the central monitoring and the local monitoring, and how to realize an intention of effective monitoring is complex and depends on user requirements and implementation possibilities.

Taking all factors into account the central monitoring level is important for multi-party business processes to collect overall monitoring information, and to arbitrate in and the resolving of conflicts. The local monitoring level is necessary for flexible adaptation to different e-market infrastructures. The local monitoring level makes it possible to keep monitorability in case the central monitoring level is not functioning properly or does not exist.

4.2   The Central Monitoring Level

Based on the above analysis, we define the central monitoring level as depicted in Figure 4. The central level consists of the following components, being a log file, contract repository, reputation repository, disputation repository, tracing module, detecting module, enforcement module, and compensation module.

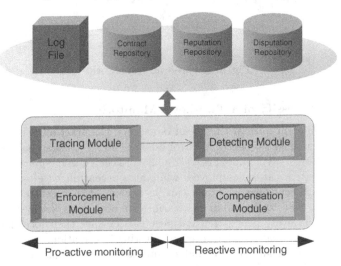

Fig. 4. Structure of central monitoring module

The functions of the three knowledge bases and a log file are as follows:

- A *contract repository* stores all agreed upon contracts that need to be executed in the e-market.
- Based on a contract from the contract repository and a *log file* in which all operations are recorded, a contract performance state can be deduced.
- A *reputation repository* records the historical status of contract fulfillments, or the use of common financial services (e.g. a bank or a credit card company) as a reputation reference.
- A *disputation repository* stores all unresolved conflicts for human-involved resolution.

The central monitoring level uses a global view to monitor an e-market. In particular in multi-party contract business processes, the central monitoring party can help the owner of the overall business process to monitor other parties which are involved but do not directly contact the main owner. For this purpose a tracing module, a detecting module, an enforcement module and a compensation module should be included to perform following functions:

- Based on the contract, the contract performance state and other monitoring information, the *tracing module* can indicate where the business process is, what needs to be done next, and signal the enforcement module after it finds non-conforming actions.
- The *enforcement module* enables the system to respond to the trace result, and to realize the intentions of avoidance or anticipation of contract violations. Reminding or warning messages can be sent to relevant parties. For non-conforming actions, an enforcement mechanism can be trigged to execute the prescribed action in accordance with the enforcement items in the contract.
- The *detecting module* ensures that any anomalous action or non-compliant action can be exactly located. It can single out and send detected results to the compensation module, whereas unsolved conflicts are sent to a disputation repository.
- The *compensation module* has the capability to compensate or undo an action that should occur, but did not due to failure.

4.3   The Local Monitoring Level

The local monitoring level has the same functions and structure as the central monitoring level. However, the local monitoring level focuses on the interests of its associated peer partner. As the local monitoring level relies solely on parties with whom it is in direct contact for its information, there are some limitations to the functions of its components. The *contract repository* only stores contracts which this local agent is involved in. The *reputation repository* can use the central reputation repository if it exists, common financial services or create local historical reputation information. The function of the *enforcement module* is limited to sending reminding and warning messages to other parties for avoidance and anticipation intentions of contract violations. In addition, the forward monitoring information module can only be used when the local monitoring level works together with a central monitoring level.

In this section, we present a two-level framework for monitoring contract fulfillment and explain reasons why a two-level framework is necessary from the e-market infrastructure, agent character aspects. Moreover, the structures of the central monitoring level and the local monitoring level are provided. For a further explanation, we will introduce a reference architecture in the next section.

5   Reference Architecture

We present a reference architecture in Figure 5 to demonstrate how our two-level monitoring framework can be applied to an agent-based central control e-market. We use a trusted third party - the *Central Monitoring Agent* - in our monitoring system. The monitorable contract will be used by all agents, either normal or central monitoring agent. All agents have the *monitoring module* and the *reactive module* [10]. The monitoring module includes a *trace module* and a *detection module*, whereas the reactive module includes an *enforcement module* and a *compensation module*.

In addition to the monitoring and reactive module, the *central monitoring agent* also has three repositories as its knowledge bases and a log file. These three repositories are: a *contract repository* to store all contracts; a *reputation repository* to record all historical status of contract fulfillments for each partners; and a *disputation repository* to store unsolvable disputations as evidence in the human-involved deal. Particularly, the central monitoring agent provides the common time standard to give each occurred action a time-stamp.

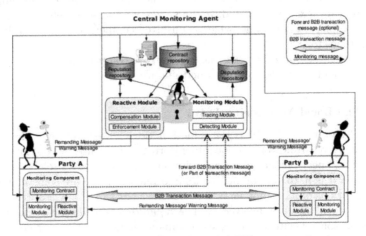

Fig. 5. The architecture for contract fulfillment monitoring

On the execution of a contract, every party can forward all or a part of the transaction messages to the central monitoring agent. It is up to the business parties to decide how much they want to be monitored. Based on these messages, the central monitoring agent identifies the particular contract from the contract repository, traces the process to identify which obligation remains on the party and the following expectable actions. Combining each partner's reputation with the current monitoring result, the central agent will send the relative reminding or warning messages, use the enforcement mechanism to enforce conforming actions, identify the necessary compensate action based on the compensation clause in the contract if any violation happened, or record any unsolvable disputations in the disputation repository.

The monitoring module and reactive module of peer agents have the same function as those of the central monitoring agent. The peer agents' reactive and monitoring module can communicate with those of the central monitoring agent to share information, and they can connect to the central knowledge bases and the log file, so that they can monitor business processes even when they did not choose the central monitoring.

The central monitoring agent/central monitoring module and peer monitoring agents/ peer monitoring module work together to provide a monitorable e-market. The two-level monitoring system shows the flexibility of the contract fulfillment being applicable from high-risk contracts to trust-based contract. High-risk contracts refer to contracts among peer business parties who do not know each other so well or simply to high business value contracts. Trust-based contracts refer to contracts among business parties who trust each other or to low business value contracts. Where the high-risk contracts need strong trusted third party monitoring, trust-based contracts may be less interested in trusted third party monitoring and instead might want to optimize on bandwidth/speed of networks.

6   Related Work

Here we review some of the main approaches to fault-tolerance in multi-agent systems as well as in the existing e-commerce platform. Hägg uses external sentinel agents to monitor inter-agent communication, build models of other agents, and take corrective actions [22]. The sentinel-based approach detects inconsistencies by observing the behavior of inter-agents. In our contract fulfillment monitoring, actions of inter-agents and external agents are all concerned with different business processes, we can extend the monitor scale from inter-agents to exter-agents because our monitoring is based on contracts.

Klein proposes to use exception-handling service to monitor the overall progress of a multi-agent system [15]. The exception-handling service is centralized approach whereas our contract fulfillment monitoring supports both centralized and decentralized monitoring.

Kaminka and Tambe use a social diagnosis approach wherein socially similar agents compare their own state with the state of other agents for detecting possible failures [3]. The socially-attentive monitoring approach is an explicit teamwork model, but it does not provide pre-active monitoring which our approach does support.

Kumar and Cohen advocate re-arranging brokers when an agent that was registered becomes unavailable [8]. This technique is implemented by adding a plan to the plan library of a generic agent. It is an efficient way for multi-agent system, but it is not realistic for contract fulfillment monitoring, since contract fulfillment monitoring is not about recovering from broker failures.

In general, multi-agent system fault-tolerant approaches analyze an entire communication going on in the system to detect state inconsistencies using replication strategies [17], sentinel approaches [22], re-assign resource approaches [8], and knowledge-based approaches [15]. Most of this research focuses on the infrastructure lever. Our contract monitoring fulfillment focuses on a semantic-level monitoring to protect an imminent contract violation.

In existing e-commerce frameworks, OASIS's ebXML provides a logging and recovery function [1], however there are no provision for fulfillment monitoring. Microsoft's BizTalk has an auditing as well as an optional document mining function [5]. Their goal is to support recovering from failures only instead of preventing them. Consequently, the development of contract fulfillment monitoring is relatively unexplored. Our framework is based a new view to solve the monitoring issues in e-market.

7   Conclusions

In this paper, we have sketched a two-level monitoring framework of e-market for monitoring contract fulfillments. Under this framework, contract violations can be prevented, whereas actual violations can be detected and compensated. With the research we have done in monitoring mechanisms [11] [12], the ultimate goal of our research is to develop a more reliable e-marketplace that behaves more similar to real-world marketplaces.

References

1. Dan A., Nguyen T.N., Dias D.M., Parr F.N., Kearney R., Sachs M.W., Lau T.C., and Shaikh H.H. Business-to-business intergration with tpaml and a business-to-business protocol framework. *IBM Systems Journal, volume 40, (no 1)*.
2. Daskalopulu A., Dimitrakos T., and Maibaum T.S.E. E-contract fulfilment and agents' attitudes. *Proceedings ERCIM WG E-Commerce Workshop on The Role of Trust in e-Business*, 2001.
3. Kaminka G. A. and Tambe M. What is wrong with us? improving robustness through social diagnosis, 1998.
4. Seaborne A., Stammers E., Casati F., Piccinelli G., and Shan M. A framework for business composition. *Position papers for the world wide web consortium (W3c workshop)*, 2001.
5. Mehta B., Levy M., Andrews G.M.T., Beckman B., Klein J., and Mital A. Biztalk service 2000 business process orchestration. *International Conference on Data Engineering (ICDE'02)*.
6. Preist C. Agent mediated electronic commerce research at hewlett packard labs, bristol. *Newsletter of the ACM special interest group on e-commerce*.
7. C. Dellarocas, M. Klein, and J.A. Rodriguez-aguilar. An exception-handling architecture for open electronic marketplaces of contract net software agents. *Proceedings of the 2nd ACM Conference on Electronic Commerce*, 2000.
8. Sanjeev K. and Cohen P.R. Towards a fault-tolerant multi-agent system architecture.

9. Weigand H. and Xu L. Contracts in e-commerce. *9th IFIP 2.6 Working Conference on Database Semantic (DS-9) Semantic Issues in E-Commerce Systems*, 2001.

10. Xu L. Research paper: Agent-based monitorable contract. 2002.

11. Xu L. Research paper: Car insurance case. 2002.

12. Xu L. and Jeusfeld M.A. Pro-active monitoring of electronic contracts. *Accepted by The 15th Conference On Advanced Information Systems Engineering (CAiSE'03)*, 2003.

13. Greunz M., Schopp B., and Stanoevska slabeva K. Supporting market transaction through xml contracting containers. *Americas Conference on Information Systems 2000*, 2000.

14. Klein M. Towards a systematic repository of knowledge about managing collaborative design conflicts. *Proceedings of the International Conference on AI in Design.*

15. Klein M. and Dallarocas C. Exception handling in agent systems, 1999.

16. Klein M. and Dellarocas C. A knowledge-based methodology for designing robust electronic markets. *ROMA Working Paper ROMA-WP-2001-02.*

17. Marin O., Sens P., Briot J., and Guessoum Z. Towards adaptive fault tolerance for distributed multi-agent systems.

18. CrossFlow Project. Insurance requirements. *CrossFlow consortium.*

19. CrossFlow Project. Insurance scenario description. *CrossFlow consortium.*

20. Jennings N.R., Faratin P., Johnson M.J., O'Brien P., and Wiegand M.E. Using intelligent agents to manage business processes. *1st Int. Conf. on The Practical Application of Intelligent Agents and Multi-Agent Technology.*

21. Angelov S. and Grefen P. B2b econtract handling - a survey of projects, papers and standards. *CTIT Technical Report 01-21; University of Twente*, 2001.

22. Hagg S. A sentinel approach to fault handling in multi-agent systems, 1996.

23. Dignum V., Weigand H., and Xu L. Agent societies: towards frameworks-based design. *The Second International Workshop on Agent-Oriented Software Engineering (AOSE-2001), Lecture Notes in Computer Science. VOL. 2222.*

24. Milosevic Z., Arnold D., and O'Connor L. Inter-enterprise contract architecture for open distributed systems: Security requirements. *WET ICE'96 Workshop on Enterprise Security.*

Simple Obligation and Right Model (SORM) - for the Runtime Management of Electronic Service Contracts

Heiko Ludwig[1] and Markus Stolze[2]

[1]IBM T.J. Watson Research Center, 19, Skyline Drive, Hawthorne, NY, 10025, USA
hludwig@us.ibm.com
[2]IBM Zurich Research Laboratory, Säumerstrasse 4, 8803 Rüschlikon, Switzerland
mrs@zurich.ibm.com

Abstract. Online purchase and delivery of goods and services requires an electronic contracting process. Formalization of contractual content enables automatic delivery of services and monitoring of the terms and conditions of the contract at service runtime. The Simple Obligation and Right Model (SORM) provides an abstract, domain-independent model of contractual content. Model instances can be interpreted and managed by applications involved in checking contractual entitlements and delivering and supervising a service in compliance with contractual rights and obligations. It captures the main types of rights and obligations and deals with their dynamics during the life-time of a contract.

1 Introduction

Electronic representations of contracts between organizations are an important prerequisite for conducting business in a networked economy, or e-business on demand. In many cases, a plain textual, natural language representation of the contractual content that can be read and interpreted by business people and lawyers suffices all requirements, now that electronic signatures stand in court in many legislations. However, if organizations want to support or automate all or a part of the life-cycle of a contract, a formal machine-interpretable representation is needed, e.g., to advertise a service, to negotiate a contract, to monitor contractual compliance, and to handle disputes [8]. This is particularly relevant in a context of Web services, which – based on standardized description and access – bear the potential to make services available across organizational boundaries on short notice for limited periods of time. A formal representation of a contract requires a model of its content, which depends on the phase of the life-cycle and the programs and people that interpret and deal with the contract.

- A model of the **structure** of a contract's content is useful in the creation and negotiation phase of a contract life-cycle. The structural elements are clauses, paragraphs, boiler plate fragments, etc. This model supports the composition process of contracts and allows word processors and other composition tools to edit and assemble contracts.
- In the fulfillment phase of the contract, we need a model of the **subject matter** of the contract, i.e. the **promises** made. A formal representation of the promises is the basis to configure, monitor and control a system fulfilling these contractual promises both from the perspective of the promising party and the

party to which is promised. A model of what is promised is also needed for consistency checking in the creation and negotiation phase.

This paper is on a model of the subject matter of the contract, its promises. A promise is directed. The party that promises enters an obligation, becoming an obligee, while the party receiving the promise becomes an obligor. The obligor holds the right that the promise will be fulfilled. Right and obligation are hence dual concepts to a promise, a different view on the same relationship.

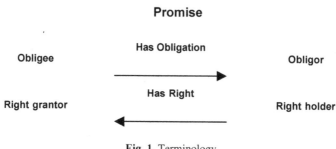

Fig. 1. Terminology

Depending on the subject domain, it is easier, or more customary, to express a contractual promise either as a right or as an obligation. For example, in contracts about the use of digital content, the promises given with respect to the digital content are usually expressed as rights that a licensing party obtains, not as an obligation of the owner of digital content to tolerate the use of the digital content by its contractual party. In the case of a contractual promise that requires one party to perform a service, this relationship is usually expressed as an obligation. Both views can be used in the same contract, e.g., stating that one party receives rights to digital contents for the obligation to pay an amount of money to the right grantor.

The design of a model of contractual promises depends on the purpose for which the model is being used. A deontic logic representation, for example, is suitable for reasoning about promises and check consistency. However, there are no obvious constructs on how and when to check entitlements for requests or when to check promises such as compliance with a response time guarantee of an electronic service. The SORM model as proposed in this paper addresses these issues. A formal specification of contractual content should be able to be interpreted according to multiple models to suit all relevant purposes.

The objective of this paper is to propose a model of obligations and rights that is independent of a particular application domain and can be used to manage obligations and rights by a contract-implementing or supervising application as well as to serve as a contribution to a contract specification language.

We proceed as follows: In the second section, we introduce an example case, derive requirements and discuss existing approaches. Subsequently, we propose a model of obligations. Section 4 introduces a model of obligation dynamics during the fulfillment of a contract. In the conclusion we give an outlook how to use the model to define formal contract languages.

2 Requirements for an Obligation and Rights Model

2.1 Example Cases

We will use two example cases to illustrate the scope of and the requirements on an obligation and rights model.

Online Services
A service provider offers to supply stock quotes for particular ticker symbols through a Web services interface. In addition, the service offers to send notifications when the market price of a particular equity either goes below or above a user-defined threshold. The quotes are offered at different levels of response time. The stock quote service provider of our example targets multiple markets: It addresses small financial agents, and actively trading consumers that only request a limited amount of quotes a day as well as enterprise customers that request large volumes. Also, customers can use the stock quote values to place them on their web site for an additional fee, i.e. Web portals are also among the target customer group. Over time, the demands of the customers may change, e.g., responding to a change of interest in more near time stock information or, in the case of financial agents, growth of the business demands more stock quotes per day. Hence, some customers want to include an option to upgrade to a higher quote volume at a price set when the original contract is made.

From the service provider organization's point of view, it enters an obligation to serve stock quotes when requested by customer, hence granting the right to request to its customer. This is bound to a particular number of requests, limiting the provider's obligation. It also guarantees a particular level of service with respect to the response time of the requests. In addition, customers may upgrade to a higher number of requests at runtime, hence modify a currently active obligation. Since the service provider addresses a heterogeneous set of customers entering in different obligations each time, it is important that the applications that offer service contracts and that monitor the services at runtime are obligation and rights-aware, addressing the particular situation of a customer. Likewise, this may be true for the service provider's customers monitoring the fulfillment of the rights they have.

Digital Content Providers
In another scenario, a distributor of music makes the music available for download over the Web. However, the downloading customer may only play the piece of music on the device onto which it has been downloaded to and may only do one backup copy. In addition, the piece of music may only be played for private use, not in public or for a paying audience.

In this example, the focus is on the rights of the customer that he or she acquires with a contract on digital music download. There are a number of restrictions to what the customer may do. However, in contrast to the previous example, the use of rights is more difficult to monitor since the online music provider is not involved when the music is actually played. The customers exercise their rights independently from a connection to the music site.

2.2 Requirements and Design Objectives

What is required to achieve the objective of an obligation and rights model that is the basis for managing those concepts in a applications implementing and supervising electronic contracts and specifying them? From the examples as well as the general understanding of contractual rights and obligations and their use in representing and processing electronic contracts, we can derive a set of general requirements in the following categories:

- **Expressiveness**; an obligation and rights model must represent all types of obligations and rights – on a sufficiently abstract level. Those main obligations and rights include the obligation to do something (send notification), the right to do something (play a piece of music) and the obligation, as well as a right, that a particular state is maintained (average response time is less than 2 seconds), each of which must be monitored in a different way. In addition, the obligation and rights model must represent how the currently active set of obligations and rights can change over the lifetime of a contractual relationship.
- **Generality**; the obligation and rights model must be independent from a particular application domain. Contract management applications, or obligation and rights-managing middleware platforms, may need to manage contracts from different application domains. However, the model should be open to refinement and extensions form particular domain.
- **Simplicity**; since the obligation and rights model must be usable across multiple domains, it is beneficial that if the model is simple, i.e. containing few concepts.

2.3 Related Work

The issue of representing the concepts of right and obligation is not novel. Models have been built for different purposes.

Deontic Logic extends first order logic (predicate logic) by modal operators with the semantics "may" and "must" [11]. Obligation and rights may be represented as expressions in this logic and may be reasoned about. Since it is based on first order logic, it very general and can be applied to any domain. This is a powerful instrument to *reason* about obligations, for example, check consistency. However, the semantics of rights and obligations is completely represented in the predicates of the logic expression, which is domain-specific. Hence, deontic logic expressions per se are not sufficient to derive how to monitor them. Furthermore, semantics of deontic logics is demanding and its concepts may not be suitable as a basis for writing applications or representing contractual content.

More recently, work on the **ODP Enterprise Language** resulted in a model of policies that cover the aspects of rights and obligations [1], [6], [12]. This model is general and expressive but deals separately with rights and obligations, which leads to redundancies. These redundancies make the model complex and prevented widespread adoption so far. Furthermore, it does not address the issue of obligation and rights dynamics.

In the area of digital content, a number of approaches to model rights have been developed. The **Open Digital Rights Language** (ODRM) can represent user rights to digital content to a great detail [5]. It is meant to be enforced by media players etc. and has a notation based on XML. Likewise, **XrML** [2] is a competing approach by another consortium. However, those approaches are restricted to their specific domains and, in general, do not address the issue of obligation.

In the field of service level agreements (SLAs), a number of approaches have been developed to formally represent the promises of a service provider. The **Web Service Level Agreement** (WSLA) approach explicitly deals with the concept of obligation, but only in the domain of SLAs [7], [8]. In addition, it does not address the concepts of rights.

In the field of **Software Engineering**, we find weaker concepts of obligation, not involving real organizations. Interface definitions such as CORBA IDL and Java interfaces may be interpreted as contracts that oblige the component implementing the interface to behave as described in its interface – an obligation to the client using the component. The approach of "design by contract" has been proposed by Meyer in the context of the Eiffel programming language [10]. However, those obligations carry very specific semantics and cannot be generalized.

3   SORM – Obligation and Right Types

In this section we introduce the obligation and right types of the Simple Obligations and Rights Model (SORM). Since all obligations and rights are derived from promises, they are directed, the right granter owing to the right holder and the obligee owing to the obligor. Abstractly, we speak of the obliged and the beneficiary. In theory, an obligation may have multiple obliged parties and beneficiaries, though this is only customary in specific areas, e.g., shared responsibility for a loan. In this case, it must be explicitly represented which parties are obliged and which parties benefit. Typically, a party in a contract is both obligor in some promises and obligee in others.

All obligations and rights have the same attributes in common. This is a unique name, the set of beneficiaries and the set of obliged parties.

The objective of SORM is to provide an obligation model that supports the monitoring of the fulfillment of obligations at runtime but is independent from a particular application domain. From the point of view of managing obligations and rights at runtime, we distinguish three basic types, depending on the way they are monitored and enforced:

- **State Obligation, State Right:** Obligation to maintain a particular state, e.g. of a service. This is relevant, for example, in service level agreements for hosting or communication service that define a service in key indicators such as through put, jitter, delay, availability etc. An obligation and right of this type must be monitored permanently, periodically, or checked each time an event occurs that may change the current state. The obligee uses the state obligation to control the fulfillment system, an obligor monitors to detect violations, which may entitle him or her to compensation (specified in another obligation).
- **Action Obligation, Right to Have an Action Performed:** Obligation of one or more parties to perform a particular action in given circumstances. This is important if a contract includes the performance of a business process or, very importantly, a payment. This is to be monitored and enforced when the

circumstances dictate that an action must be performed or the performance must be completed. The performance of the action can be scheduled by the obligee and the monitoring of its correct completion can be scheduled by the obligor.

- **Option Obligation, Right to Act:** Obligation to tolerate an action performed by another party. This includes making a service operation available, e.g., to retrieve equity quotes, and to allow a client to invoke it. This also includes the right to play digital content. From the point of view of the option obligation's beneficiary, it represents a right for this party to perform an action. It is solely up to the right holder when to execute an action, e.g., a user starts playing a piece of music on his or her device. The entitlement to an action subject to the right must be checked upon invocation but cannot be scheduled. In the case of electronic services, the management of rights to act is associated with access control.

Each of these types can be subtyped if appropriate for a particular application domain.

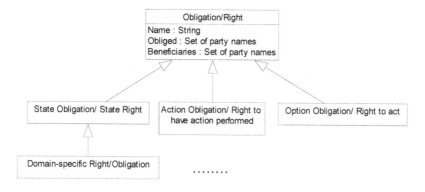

Fig. 2. Obligation and right types

In the subsequent discussion, we refer to rights and obligations mainly in their obligation form to keep it simple. Due to the duality of right and obligation, the corresponding right view is implied.

3.1 State Obligation, State Right

A state obligation guarantees that a particular condition holds in a defined period. We define a state obligation as a tuple:

so(name, OP, BP, c, period), where,

- *name* is the unique name of the obligation,
- *OP* is a set of obliged parties,
- *BP* is a set of beneficiary parties,
- *c* is the condition that must hold, representing a state of the world, and
- *period* is the period of time in which the condition must hold.

The following example is drawn from the case of section 2:

so(so1,
          {Stock quote service provider},
          {Stock quote customer},
          a 95 percentile of stock quote requests, measured on an hourly basis, are re-
          turned to in 3 seconds,
          runtime of the contract)

In this simple example, the service provider promises to the customer a particular response time behavior throughout the runtime of the contract. In the given example, the guaranteed condition cannot be related to each request individually. Therefore, this obligation may be defined as a state obligation in the way shown above.

Another example relating to the class of service of the stock quotes:

so(so2,
          {Stock quote service provider},
          {Stock quote customer},
          the returned stock quotes are the current NYSE market values delayed no
                more than ten minutes,
          runtime of the contract)

This example refers to the delay of the quoted values with respect to the market price setting, not the technical delay as in the first example.

3.2   Action Obligation, Right to Have Action Performed

An action obligation expresses that the obliged parties must perform an action a if and when a given event occurs and defined preconditions are met. We define an action obligation as a tuple:
ao(name, OP, BP, a, e, pre, completiontime), where

- *name* is the unique name of the obligation,
- *OP* is a set of obliged parties,
- *BP* is a set of beneficiary parties,
- *a* is an action to be performed, potentially a complex, structured process comprising multiple steps,
- *e* is an event on whose occurrence the action is to be performed, if the precondition holds,
- *pre* is the precondition that must be met for the action to be executed at event e,
- *completiontime* is the specification of the point in time until the action must be completed. This specification can be either relative to the start of the action or an absolute point in time.

Again, we take an example from the stock quote service case:

ao(ao1,
          {Stock quote service provider},
          {Stock quote customer},
          send instant message notification about price of IBM stock and then debit 10
          cents to service customer's account ,

new deal closed on IBM stock at NYSE,
price of deal more than $130 per share,
1 minute after the deal was published by NYSE)

In this example, the stock quote service provider promises to the customer to send an instant message containing the stock price of IBM within 1 minute if and when a deal of more than $130 per share is closed on NYSE. The precondition must be evaluated each time a deal on IBM stock is closed on NYSE. For this service 10 cents will be debited to the customer's account with the service provider.

The example is a promise to perform a particular action and thus is well represented as an action obligation.

3.3 Option Obligation, Right to Act

An option obligation, or right to act, expresses that the obliged parties allow the beneficiary parties to perform an action. This only makes sense if the beneficiary parties are not entitled to so anyway, even if no contract is signed. Also, the obliged parties must be able to grant this right to the beneficiaries.

We define an option obligation as a tuple:
oo(name, OP, BP, a, pre), where

- *name* is the unique name of the obligation,
- *OP* is a set of obliged parties,
- *BP* is a set of beneficiary parties, the parties that may perform the action,
- *a* is the action that the members of BP may execute, which may be complex,
- *pre* is the precondition that must be met for the action can be performed.

This type of obligation is illustrated in the following example:

oo(oo1,
 {Digital content distributor},
 {Digital content customer},
 play digital content,
 weekdays,
)

The above example represents the main service function of our case. The digital content customer may play digital content only weekdays.

Another example:

oo(oo2,
 {Stock quote service provider},
 {Stock quote customer},
 (set new stock quote delay from 10 minutes to 5 minutes,
 debit $ 10000 to service customer's account)
 time of request is between 1 AM and 1 PM EST,
)

In this example, the service customer may improve his or her class of service in terms of the time delay of the stock quote. While the original contract foresaw to return quotes as of 10 minutes ago, the customer can decrease the delay to 5 minutes, for a price of $10000.

3.4 Representation of Obligation and Right Content

The above definitions and examples of obligation and right types outline the structure of these constructs. However, to be able to automatically process contracts we need a formal representation of the content of the obligation and right elements, too. There are multiple ways to formalize the content, i.e. the conditions, events, time expressions, and the invocation of activities. None of these aspects is novel and suitable formal representations can be used where appropriate, depending on the particular domain. However, the contract must capture the obligation content in a way that is mutually understood by the contracting parties, based on a shared ontology. We need formal representations for the following elements of obligations and rights:

Conditions can be interpreted as Boolean functions that can be resolved to true or false at runtime. To be automatically verifiable, condition must be expressed as expressions in a formal language. In this formalism, we may use first order logic or a modal logic where appropriate. The logic variables used in the predicates are unique names of objects defined in the context of the obligations, i.e. the contract. The choice of the appropriate set of predicates to be used depends on the particular domain of the contract. Both objects and predicates must be based on a shared ontology of the contracting parties.

Example:

$$greater(last\_ibm\_price, 130)$$

The semantics of the object *last_ibm_price*, being the share price, and the predicate *greater* must be understood between the parties.

Points in time are either absolute date and time descriptions or they are relative to a given point in time, e.g., 60 seconds after the *getQuote* operation was received. Periods are defined by their start and end point in time. Any appropriate formal representation fits.

In action obligations we define which **action** must be performed in case the precondition holds. We assume a model of action that is similar to a function in a programming language, having a name and a set of parameters. Each reference to perform an action thus includes (1) the name of the action and (2) the marshalling of its parameters. The parameters are either names of objects of the contract or constants.

Examples:

$$set(stockQuoteDelay, 10)$$
$$debit(customerAccount, 10000)$$

In these examples, stockQuoteDelay and customerAccount are names of objects of the contract. In some cases, actions are not atomic but it is agreed to perform a process consisting of multiple actions in an execution order. These action descriptions can be implemented by descriptions of simple invocations of a Web service or a process, e.g., represented in BPEL 3[3].

We require a formal **event model**. Simple events could be defined by their unique name. If necessary, this model can be extended to complex, structured events.

4 SORM - Obligation and Right Dynamics

Within the lifetime of a contractual relationship between two parties, the **set of obligations** and the **currently valid set of obligations** can vary.

New obligations can be introduced by one contracting party exercising an option obligation. For example, a party may choose a higher quality of service level, which entails that there arises the obligation to pay an additional fee. Some option obligations are designed to be exercisable an indefinite number of times, hence introducing an arbitrary number of new obligations, depending on the beneficiary of the option obligation. Also, executing options can also lead to a reduction of the set of obligations, e.g., by replacing obligations with new ones or simple removing obligations. Note that the expansion of the obligation set of a contract at runtime is always prearranged.

Each obligation can be currently valid or not. An obligation is valid if its precondition is true but the obligation has not been fulfilled yet.

This section introduces a model of managing the dynamics of the set of obligations at runtime.

4.1 Actions Modifying the Obligation Set

Changing the set of valid obligations requires action to do so. We introduce an action type that adds a new obligation:

addObligation(obligation): Obligations := Obligations ∪ {obligation}, where

- *Obligations* is a set of Obligations of a contract and
- *obligation* is an individual obligation.
 ":=" represents an assignment operator similar to programming languages. Likewise, we define an action that removes an obligation from the current set:
 removeObligation(obligation): Obligations := Obligations / obligation, where
- *Obligations* is a set of Obligations of a contract and
- *obligation* is an individual obligation to be removed from the set.

For convenience reasons, we introduce a changeObligations operation that adds and removes sets of operations with one operation.

changeObligations(ObligationsRemoved, ObligationsAdded):
        Obligations := (Obligations / ObligationRemoved) ∪ ObligationsAdded,
where

- *ObligationsRemoved* is a set of Obligations of a contract that will be removed from the current set of obligations and
- *ObligationsAdded* is the set of obligations that is added to the curre nt set.

These operations can be used in obligations like any other action.

4.2 Using Modifying Operations in Obligations and Rights

In many cases, actions executed in the context of action obligations or rights to act entail a change in the set of currently active obligations. In those cases, the addObligation, removeObligation, and changeObligations operations are used in action obligations and rights to act. This is the case, for example, in the case of our stock quote service providing its customers the option to invoke the stock quote operation. If the customer exercises the option, a new obligation arises for the provider, namely to perform an action under given conditions. This is just a new action obligation. The following example illustrates this:

```
oo(    getStockQuote,
       {Stock quote service provider},
       {Stock quote customer},
       addObligation (  ao(  ao2,
                             {Stock quote service provider},
                             {Stock quote customer},
                              return price of the requested stock,
                             on activation,
                             <no precondition>,
                             within 2 seconds
                        )
                     ),
       time of request is between 9 AM and 5 PM EST,
    )
```

This right to act "getStockQuote" grants the customer the right to activate a new action obligation that obliges the provider to execute the action "return price of the requested stock". This right is available between 9 AM and 5PM. The action must be executed immediately with no further pre-condition and must finish within 2 seconds. The service provider may charge $1 for each quote. In this case, two action obligations are added, the one above and an additional one obliging the customer to pay $1, for example, by the end of the month.

```
oo(    getStockQuoteForMoney,
       {Stock quote service provider},
       {Stock quote customer},
       changeObligations (        {},
                                  { ao(  ao2,
                                         {Stock quote service provider},
                                         {Stock quote customer},
                                         return price of the requested stock,
                                         on activation,
                                         <no precondition>,
                                         within 2 seconds
                                    )
                                  ),
                                  ao(  payment,
                                       {Stock quote customer},
                                       {Stock quote service provider},
```

pay one dollar,
on activation,
<no precondition>,
until the end of the month
)

}

),

time of request is between 9 AM and 5 PM EST,

)

The first parameter of the changeObligations operation is the empty set since we only added obligation.

Many contracts include the right to upgrade to a different level of service, which represents a change in a state obligation in our model. Also, we find scheduled changes of the obligation set, such as in the following example:

ao(scheduledQoSUpgrade,
{Stock quote service provider, Stock quote customer},
{Stock quote customer, Stock quote service provider },
changeObligations ({ so(95percentileQoS,
{Stock quote service provider},
{Stock quote customer},
a 95 percentile of stock quote requests,
measured on an hourly basis, are re
turned to in 3 seconds, runtime of the
contract)

},
{ so(98percentileQoS,
{Stock quote service provider},
{Stock quote customer},
a 98 percentile of stock quote requests,
measured on an hourly basis, are re
turned to in 3 seconds, runtime of the
contract)

}),

August 1ˢᵗ 2004,
<no precondition>,
before 8 AM on August 1ˢᵗ 2004)

In this scheduled QoS upgrade, the parties agree to upgrade the state obligation defining the percentile of requests served within 3 seconds on August 1ˢᵗ 2004. This means removing the old state obligation defining the 95 percentile requirement of the provider and adding the new state obligation. Since this change is scheduled, and hence an action obligation, it can be seen that both parties must implement this change of status and both parties benefit from it.

4.3 Obligation and Right States

In large sets of obligations and rights, it is inconvenient to define large modifications of the current state set for each action that may cause it. In addition, the dynamics of large sets of obligations are difficult to manage on an obligation-by-obligation basis. We need an abstraction mechanism that allows us to group obligations and manage entire obligation groups.

We introduce the notion of an obligation state. An obligation state is a tuple: *orstate(Obligations)*, where

- *Obligations* is a set of obligations and rights that are valid when obligation and rights state (OR state) is activated.

Once the OR state is activated, smaller changes to the obligation set can be caused by *addObligation* and *removeObligation* actions.

We need a new action to change to a new obligation state:
newState(state): CurrentObligations := NewObligations$_{state}$, where

- *state* is the state that the contract is to be set to,
- CurrentObligations is the current set of obligations and rights,
- *NewObligations$_{state}$* is the new set of obligations and rights, corresponding to the state.

The current set of obligations and rights is replaced by the new set.

4.4 Consolidating Obligation and Right States with Individual Dynamics

The combined use of OR states and individual changes of the obligation and rights set based on option obligations can lead to ambiguities about the currently valid set of obligations. What happens to a newly added obligation when OR states are switched? For this purpose, we distinguish two sets of obligations and rights in a contract.

- **Background obligations and rights** are obligations and rights that are in force independently of the current OR state.
- **State-based obligations** are the obligations and rights of the currently active OR state.

Individual add and remove operations on obligations and rights only modify the background obligations and rights, which are not affected by changes in the OR state.

The set of obligations of a contract comprises the background and the state-based obligations.

The figure illustrates the use of background obligations and rights and OR states. In this example, the current set of background obligations and rights comprises state obligation 1, action obligations 2 and 3, and option obligation 4. This set of background obligations can be modified using the addObligation and removeObligation actions (in blue). In addition, the contract comprises three obligation states. States are changed using the newState action, which, for example, could be offered as an option obligation or as an action that is to be performed at a given time. Currently, OR state

2 is active. Hence, the currently valid set of obligations comprises the background obligations plus the obligations of OR state 2.

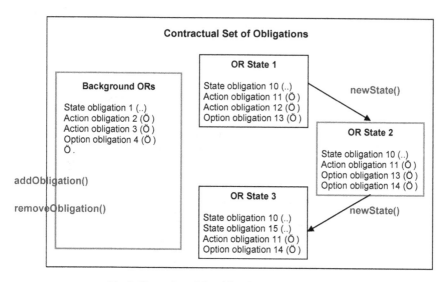

Fig. 3. Dynamics of the obligation set of a contract

5 Summary and Conclusion

This paper proposes the Simple Obligations and Rights Model (SORM) as a domain-independent basis for the representation and management of promises in applications implementing and supervising electronic contracts This is particularly relevant in a Web services context where business relationships may be set up at short notice. Based on the duality of obligations and rights, the model can express – abstractly – all forms of obligations and rights based on three types, state obligation - right, action obligation – right to have action performed, and option obligation – right to act, distinguishing the different ways these promises are monitored and enforced at runtime. Using operations modifying the set of currently active obligations and rights, the model can also capture the dynamics of obligations and rights over time, hence facilitating long-running and complex contracts. SORM can be used as a basis for applications dealing with rights and obligations beyond a particular application domain.

SORM can also be used to design a formal contract language. Obligations and rights are primarily used in the context of contracts between organizations or individuals. The expressiveness of the contract language, i.e. which subset of SORM is actually used, and the specific syntax must depend on the capability of the contract-interpreting application to deal with the flexibility of potential obligation and right expressions. If an application can only deal with, say, state obligations, the language can be limited to this type. Also, a contract language must address to formalize descriptions of conditions, events, actions and times, to the extent it needs. SORM has been developed as a generalization of the Web Service Level Agreement (WSLA) language in which many ideas of SORM have been tested in an environment of SLA management [7], [9].

Since the obligations and rights model of a contract language and a contract management application should match, SORM is a good candidate for a contributor to the overall content model of an electronic contract.

References

1. J. Cole, J. Derrick, Z. Milosevic, and K. Raymond: *Policies in an enterprise specification*. In Morris Sloman, editor, *Proceedings of the Policy Workshop*, 2001, Bristol UK, January 2001.
2. ContentGuard : eXtensible rights Markup Language (XrML) Specficiation 2.0 – Part 1: Primer. 20 November 2001. http://www.xrml.org/ on March 27, 2003.
3. F. Curbera, Y. Goland, J. Klein, F. Leyman, D. Roller, S. Thatte, S. Weerawarana; *Business Process Execution Language for Web Services (BPEL4WS) 1.0*; August 2002, http://www.ibm.com/developerworks/library/ws-bpel
4. Y. Hoffner, S. Field, P. Grefen, H. Ludwig: Contract-driven creation and operation of virtual enterprises. In *Computer Networks 37*, pp. 111–136, Elsevier Science B.V. 2001.
5. R. Iannella: *Open Digital Rights Language (ODRL)*. W3C Note, 19. September 2002, http://www.w3.org/TR/odrl/.
6. ISO/IEC JTC 1/SC 7: Information Technology - Open Distributed Processing - Reference Model - Enterprise Language: ISO/IEC 15414 I ITU-T Recommendation X.911,. Committee Draft. 8. July 1999.
7. A. Keller, H. Ludwig: The WSLA Framework – Specifying and Monitoring Service Level Agreements for Web Services. *Journal of Network and System Management* (11), Nr. 1, Special Issue on E-Business Management. Plenum Publishing Corporation, 2003.
8. Ludwig, H; Hoffner, Y: The Role of Contract and Component Semantics in Dynamic E-Contract Enactment Configuration. In *Proceedings of the 9th IFIP Workshop on Data Semantics (DS9)*, pp. 26–40, Hong Kong, 2001.
9. H. Ludwig, A. Keller, A. Dan, R. King, R. Franck: A Service Level Agreement Language for Dynamic Electronic Services. Electronic Commerce Research (3), Nr. 1, pp. 43–59, Kluwer Academic Publishers, 2003.
10. B. Meyer: Object-oriented Software Construction. 2nd ed. Prentice-Hall, 1997.
11. J.-J.Ch. Meyer and R.J. Wieringa (eds.): *Deontic Logic in Computer Science: Normative System Specification*. Wiley and Sons, 1993.
12. M. W. A. Steen, J. Derrick: ODP Enterprise Viewpoint Specification. *Computer Standards and Interfaces*, 22:165–189, September 2000.

Integrating Context in Modelling for Web Information Systems[*]

Roland Kaschek[1], Klaus-Dieter Schewe[1],
Bernhard Thalheim[2], and Lei Zhang[1]

[1] Massey University, Department of Information Systems
& Information Science Research Centre
Private Bag 11 222, Palmerston North, New Zealand
[r.h.kaschek|k.d.schewe]@massey.ac.nz, maggiezl@hotmail.com
[2] BTU Cottbus, Department of Computer Science
Universitätsplatz 1, 03044 Cottbus, Germany
thalheim@informatik.tu-cottbus.de

Abstract. While using large-scale Web Information Systems (WISs) customers sometimes get confused and experience difficulties in solving their business problems. We propose context modelling as a conceptual tool for aiding customers to regain the possibility of using the respective WIS efficiently. We integrate context modelling into the co-design methodology, identify some losing track patterns and propose respective customer aid. Finally we formalise our approach based on the theory of media types and the model of contextual information systems developed by Teodorakis et al.

1 Introduction

Traditional definitions of the term "Information System" (IS) emphasise that such systems have to serve a given business purpose. In doing so they allow customers to record, store, infer and disseminate linguistic expressions [8]. According to the work by Mayr et al. [12] we use the metaphor of information space created by an IS to organize the linguistic expressions that are stored, disseminated or derived with help of the IS. This information space consists of locations at which information objects are located and connections between them. Users can navigate through this space, locate data and functionality, invoke functionality, filter, reorder, reshape, and export or import data from or to other integrating spaces. Furthermore users may enter or leave information spaces.

To be able to cope with the complexity of the information space of a traditional, i.e., non-Web information system users are usually involved in requirements analysis and design and trained in using realised systems. However, if the

[*] The research reported in this article is supported by the Massey University Academy of Business Research Fund (Project "Story boarding for web-based services", 2002/03) and the Information Science Research Centre.

functionality of the IS is made available to an unknown world-wide community of users by providing a web interface, thus turning it into a "Web Information System" (WIS), customer involvement is more or less impossible. As a consequence, customers may get lost in the information space or confused by its complexity. We claim that a well-designed WIS must cope with these problems in the WIS itself. Therefore, we will need tools that help who got confused or lost.

Providing aid to such customers requires a thorough understanding of what these customers actually do. This involves having a model of the context of the respective usage situation as part of the WIS. We propose defining a separate context space to associate with each actual usage situation a point in the context space. We thus propose to provide a mapping from situations to contexts. Furthermore, we require that each context provides a set of means that aid customers to regain awareness of how to proceed for solving their business problem.

Related Work. In [2] three inter-related aspects of WISs have been presented emphasising *content*, *navigation*, and *presentation*, which lead to databases, hypertext structures, and web page structures, respectively. Several authors such as [3,4,10,14,17] have more or less followed this approach to WIS conceptualisation and contributed to integrated methodologies for WIS development. However, the usage has almost been neglected, and functionality has very often been reduced to just navigation.

The co-design methodology has been presented in [16,15]. It employs story boarding to cover the usuage aspect — this has been dealt with in detail in [6,15] — and the theory of media types as its formal foundations [9]. These foundations have been linked to XML in [11]. Application experiences concerning co-design are reported in [7,20].

We believe that despite its merits co-design does not yet focus on all aspects of context modelling nor does it provide optimal support for the weak form of of context modelling, which it currently focusses on. Our goal is to enhance co-design in these respects. In [21,7], respectively, the terms "localization abstraction" and "escort information" have been used to address context modelling issues. In [9] an approach to context modelling is proposed, which is based on a subtype relationship between media types. This approach, however, is a static one, as it associates with each location L in an information space a set of linguistic expressions that do not depend on the path that led a customer to the location L. This static concept is useful for tree-like navigation structures. However, for more complex navigation structures a dynamic solution is required reflecting the various possible paths taking a customer to a location L.

Our Approach to Context Modelling. We exploit the work on contextual information bases (CIB) [1,18,19] for enhancing the co-design methodology with context modelling. Roughly speaking, a CIB-context associates objects with a name and an optional reference to another CIB-context. Thus, both the name and the reference depend on the usage history. To exploit CIB-contexts we slightly generalise them. Instead of only associating merely a name with a location L we associate a location with it, i.e. a complex value.

We do not discuss other forms of contexts such as the real world venue context, i.e. the history of venues at which the customer interacted with the IS. Furthermore, in the sequel we only comment on the relevance for information systems development and ignore system test or validation. More profound discussion of various aspects of modelling and using contexts can be found for instance in the Context'99 proceedings [5].

Paper Outline. We proceed with a brief introduction of the abstraction layer model and the co-design methodology in Section 2. We continue with a discussion of context and context modelling in Section 3. A presentation of our formalisation of contexts on the basis of the theory of media types and the model of contextual information systems follows in Section 4. We finish the paper with a brief outlook.

2   Abstraction Layer Model and Co-design Methodology

We briefly describe core parts of the Abstraction Layer Model (ALM) and the Co-design Methodology (CDM) for the design and development of WISs.

The ALM identifies layers of abstraction of IS concepts. The *strategic layer* addresses the purpose of the WIS, i.e. its mission statement and the anticipated customer types including their goals. The next lower layer is called the *business layer*. It deals with modelling the anticipated usage of the WIS in terms of customer types, locations of the information space, transitions between them, and exchange of linguistic expressions between these locations.

The central layer is the *conceptual layer*. Its purpose is to organise the linguistic expressions in a reasonable way. On this abstraction layer the separation of concerns that was introduced at the business layer by modelling the system usage from a customer type perspective at this abstraction layer is carried over into an integrated conceptual model. The locations identified at the business layer are turned into so-called media types, i.e. units of data and functionality. The integrated collection of media types forms the conceptual model. The media types are linked to an underlying database by means of views, which permit the derivation of data and functionality that must be made accessible to customers at the respective location.

The next lower layer is the *presentation layer*. It is devoted to the problem of allocating access channels to the media types. This can be seen as a step towards implementing the system. The lowest layer is the *implementation layer*. All the aspects of the physical implementation have to be addressed on this layer, including setting up the logical and physical database schemata, the page layout, the realisation of functionality using scripting languages, etc.

Except for the strategic layer the dimensions *modus* (static, dynamic) and *focus* (local, global) apply to the linguistic expressions relevant for the WIS. Therefore, throughout the development of WIS CDM requires 'data' (static &

global), 'view' (static & local), 'functionality' (dynamic & global), and 'dialogue' (dynamic & local) to be considered at all abstraction layers except the strategic layer. Following this guidance leads to a development process of WIS, which does not only focus on the most important aspects of WIS but also organises the development process in a stepwise manner that takes care of the most important development artifacts.

3   Towards Context Modelling

According to the Langenscheidt-Longman Dictionary of Contemporary English the *context of something* is 'the situation, events, or information that are related ...[to this particular something] ...that help you to understand it better'. Our problem here is to understand what a customer did while accessing a particular WIS. This includes his or her actual goals and if these were not achieved the reasons for it. We focus on how to support a customer in solving his or her business problems. This task requires a customer model being represented in the WIS, analysed and used for determining appropriate aid, while the customer is operating the WIS. We consider it an important feature of a WIS that access to important parts of the offered services is unrestricted.

ALM and CDM propose identifying customer types to introduce a usage-based separation of concerns as the basis of WIS development. These customer types are considered as the core of the customer model represented in the WIS and are used for determining customer aid in losing-track situations, i.e. situations where customers seem not following a plan of solving a problem efficiently.

3.1   Context Spaces

We introduce the concept of *context space* as our key conceptual tool for context modelling. In a losing-track situation the actual WIS state is to be mapped onto a point in the context space, which then is used to determine the best suitable means that helps the customer overcoming the losing-track situation.

We approach the problem of providing customer aid by introducing subspaces of the information space according to business peculiarities. Additionally the customer is asked to identify the subspace she or he wants to access. This approach is used by a number of WISs. In slightly more theoretical terms it can be based on customer types, which refer to typical ways of doing the business the WIS is about. The provision of such customer types encourages (up to some degree) the customers to supply correct and precise data, for the customer must expect the WIS not being able to support her or him well in case she or he supplies false or imprecise data.

Our restriction to an approach to context modelling focussing on a business model is debatable. Sometimes customers might want to use a WIS for a purpose that does not meet the system's mission statement. For example, a customer might want to use a bank WIS for learning about the loan business, or a bookshop WIS for learning English. Clearly, the larger the gap between the actual customer's intention and the system's mission statement is, the higher the expected costs will be for supporting such customers. If it can be forecasted that some customers will interact with the WIS in a 'non-standard' way, a decision might be made to support such intentions or not. This implies a modification of the anticipated information space. It shows that our focus on a business model for context modelling is not a severe restriction.

According to CDM a customer type there is associated with a list of customer characteristics, each of which is endowed with a linearly ordered scale of values. For instance, customer characteristics for a bank WIS might be age, education, or assets, as banks might offer business models suitable for individuals of a particular age (e.g. under 12, between 12 and 18, over 65); of particular education (e.g. students might be given convenient conditions in running accounts, as they are expected to have higher incomes after graduation); of assets (e.g. wealthy customers might be interested in buying shares and depositing larger amounts of money with the bank). The customer type characteristics — if chosen to be independent — generate a multi-dimensional space. Convex subsets within it according to CDM may be defined as *customer types.*

The association of a customer type to a customer at least in part is a dynamic association. This is a consequence of binding customer types to particular business models and information subspaces. Clearly, some customer characteristics such as gender, ethnicity, age and name are expected not to change throughout a WIS-session. Other customer characteristics, however, might change throughout a session. These can be expected to express the actual customer intent. A re-association of a customer type to a customer of course should only be carried out after explicit permission to do so has been given by the respective customer.

Customer types may be considered as being particular views on the use the customer makes of the WIS, because a part of the customer interactions is used to classify the respective customer as being of a certain type. Occasionally this particular view may not be sufficient to determine suitable customer aid. Therefore, we add the usage history dimension to the space generated from the union of the sets of customer-specific characteristics to define the context space. The usage dimension of the context space is the set of all paths through the information space created by the WIS.

A view V must be made accessible to the customer. Thus, a *view* $V = V(T, L, H_L)$ depends on the customer type T, the usage history H_L, and location L in information space that the customer has reached after H_L. We propose to organise the ongoing customer support in three steps:

- *Abstracting* the view, i.e. determining for a customer in location L the view $V(T, L, H_L)$ on the history of his or her usage of the WIS.
- *Detecting* whether a losing-track situation has occurred, i.e. continuously checking the history H_L for each reached location L in the information space.
- *Proposing* the view $V(T, L, H_L)$ that most likely will enable him or her to continue problem solving, i.e. making this view accessible to the customer in case a losing-track situation was detected.

In order to guarantee that the history H_L can be obtained for all locations L, a session based activity log of the customer can be used. For each visited location L a reference to the location L' visited immediately before L and a record of the interaction I' that occurred in L' is added to the log. The view $V(T, L, H_L)$ must then be computed from the list of all locations $L_n, \ldots, L_1$ that the customer has visited within a reported time interval Δ, where $L_n = L$, and L_1 is the first location visited within Δ.

3.2   Losing-Track Situations

Losing-track situations can be detected based on the customer's behaviour, e.g.:

- invoking the help function repeatedly on similar topics;
- repeatedly positioning on particular locations in information space and performing similar operations on similar data;
- excessively navigating through information space without accessing data;
- looking repeatedly for FAQs on similar topics;
- attempting to enter a discussion forum;
- sending email to the site administrator.

Customer aid that can be provided for losing-track situations is giving access to a thesaurus of the information subspace the customer is accessing. Furthermore, the respective business model may be exposed to the customer together with an explanation that is adapted to the customer type the customer conforms to. Similarly, access to a FAQ list suitable for the customer type and the accessed information subspace may be given. Furthermore, improved search facilities and examples targeting at the information subspace accessed might be provided.

Besides these generic ideas for customer aid more specific ones can be generated based on two complementary ideas. Firstly, story boarding must not be limited to a forward-engineering approach, which models how a customer can achieve his or her goal. Additionally, a backward-engineering approach is required, which models losing-track situations that are likely to occur and the aid that could be offered to a customer to overcome such situations.

Secondly, customer responses due to losing-track situations while the WIS is effective should be collected, systematised, classified and made the starting point of maintenance and redesign tasks. Thus, observing the respective customer-supplied data is of particular importance for providing highly customer- and situation-specific aid.

4   A Formal Concept of Context

We now address a more formal approach to context modelling based on an extension to the theory of media types [9,16]. We briefly describe the fundamentals of media types being generalised views over an underlying database, then we take a look at contextual information bases (CIBs) [18,19], a separate formalism for modelling contexts. We then generalise and tailor the CIB-approach in order to integrate it with the theory of media types. This provides the formal conceptual means for context modelling as outlined in the previous section.

4.1   Media Types

The core of a media type is defined by a view. A *view* V on a database schema S consists of a view schema S_V and a defining query q_V, which transforms databases over S into databases over S_V.

The underlying datamodel itself is not relevant. The defining query may be expressed in any suitable query language, e.g. query algebra, logic or an SQL-variant, provided that the queries are able to create links, i.e., the used query language must have the *create-facility* [9].

Media types already provide a limited form of modelling contexts, which is realised by a supertyping mechanism. This leads to the definition of *raw media type* based on some type system. The type system must provide base types and type constructors, e.g. record, set and list type constructors. Arbitrary *type expressions* are built by nesting these constructors.

A *raw media type* has a name M and consists of a content data type $cont(M)$, which is an extended type expression, in which the place of a base type may be occupied by a pair $\ell : M'$ with a label ℓ and the name M' of a raw media type, a finite set $sup(M)$ of raw media type names M_i, each of which will be called a supertype of M, and a defining query q_M with create-facility such that $(\{t_M\}, q_M)$ defines a view. Here t_M is the type arising from $cont(M)$ by substitution of URL for all pairs $\ell : M'$.

In order to model functionality operations are added to raw media types. An *operation* on a raw media type M consists of an operation signature, i.e., name, input-parameters and output-parameters, a selection type which is a supertype of $cont(M)$, and a body which is defined via operations accessing the underlying database.

In order to allow the information content to be tailored to specific customer needs and presentation restrictions, raw media types are extended to media types. The most relevant extension is *cohesion*, which introduces a controlled form of information loss. Formally, we define a partial order $\leq$ on content data types, which extends subtyping in a straightforward way such that references and superclasses are taken into consideration.

If $cont(M)$ is the content data type of a raw media type M and $sup(cont(M))$ is the set of all content expressions exp with $cont(M) \leq exp$, then a total pre-order $\preceq_M$ on $sup(cont(M))$ extending the order $\leq$ on content expressions is called an *cohesion pre-order*. Clearly, $cont(M)$ is minimal with respect to $\preceq_M$.

Small elements in $sup(cont(M))$ with respect to $\preceq_M$ define information to be kept together, if possible. An alternative to cohesion preorders is to use *proximity values*, but we will not consider them here.

A *media type* raw media type M extended by operations and a cohesion pre-order $\preceq_M$. Details on the theory of media types have been published in [9,15].

In this work further extensions beyond cohesion are discussed, but these are not relevant for context modelling.

We defined that part of a media type M is a finite set $sup(M)$ of media type names M_i $(i = 1, \ldots, n)$, each of which will be called a supertype of M. Thus, if the defining queries are evaluated for M and all the M_i $(i = 1, \ldots, n)$, we obtain complex values v and v_i $(i = 1, \ldots, n)$ for the types t_M and t_{M_i} $(i = 1, \ldots, n)$, respectively. The collection of these values together with a generated URL u defines a media object of type M. In this media object the values v_i $(i = 1, \ldots, n)$ can be considered as context information, as they do not depend just on M. Furthermore, these values v_i may also appear in other media objects.

4.2   A Concept of Context

Using supertypes for context modelling is only static. As soon as there are several paths leading to a particular scene sc associated with the media type M, the context information provided by the supertypes of M must refer to all these paths at the same time. If we want to provide path-specific context information, this will turn out to be insufficient.

Therefore, we have to investigate alternatives. The theory of contextual information bases developed by Teodorakis et al. [18,19] seems to be a promising approach to be considered. According to this theory a context is a set of objects, each having several names, and each of these names may be coupled with a reference to another context. There may be names for objects that are not referencing to other contexts. Here, the term 'object' is used in the sense of 'object identifier', i.e. a unique abstract handle to identify objects.

More formally, a *context* C is a finite set of triples (o_i, n_i, r_i), where o_i is an object identifier, i.e. a value of some base type ID, n_i is a name, i.e. a value of type $STRING$, and r_i is either a reference $\rightarrow C'$ to a context C' or nil, the latter one indicating that there is no such reference.

We write $C = \{n_1 : o_1 \rightarrow C_1, \ldots, n_\ell : o_\ell \rightarrow C_\ell\}$. If there is no reference for the i'th name, i.e. we have (o_i, n_i, nil) we simply omit $\rightarrow C_i$ and write $n_i : o_i$.

The idea of working with contextual information bases is that a customer queries them and thus retrieves objects. In order to describe these objects in more details she or he accesses the context(s) of the object, which will lead — by following the references — to other objects. In addition, a particular information encoded by the name is associated with each of these references. The work in [1] describes a computer animation tool for working with such contextual information bases.

The work in [19] describes a path query language for contextual information bases. Most important for our problem are the following macros of this language:

- The macro look-up(C, n) takes two input parameters. The first one is the name of a context C. The second one is a name n, i.e., a value of type $STRING$. The macro returns name paths $n_i = n_i^0, \ldots, n_i^{k_i}$ starting from context C and ending in n, i.e. $n_i^{k_i} = n$.

- The macro **cross-ref**(p, C) also takes two input parameters. Here the first one is a name path $p = p^0, \ldots, p^\ell$. The second one is the name of a context C. The macro returns name paths $n_i = n_i^0, \ldots, n_i^{k_i}$ starting from context C and ending in the name specified by p, i.e., $p^\ell = n_i^{k_i}$.

4.3   Combining Media Types with Contexts

Let us now bring together media types and contextual information bases. The obvious questions are:

- What are the objects that are required in contextual information bases, if we are given media types?
- What are the references that are required in contextual information bases?
- Is it sufficient to have names for describing objects in a context or should these be replaced by something else?

The natural idea for generalising the notion of object in contexts is to choose the media objects. Concentrate on the raw media objects first. Evaluating the defining query for a raw media type M leads to a set $\{(u_1, v_1), \ldots, (u_n, v_n)\}$ of raw media objects. Recall that the u_i are values of type URL, whereas the v_i are values of the representing type t_M. As these URL-values are unique, they identify the raw media objects, and thus can be used as surrogates for them in the notion of context.

This answers our first question. The objects are the (raw) media objects. The object identifiers needed in the contexts are the (abstract) URLs of these media objects.

As we want to have access to path information, we may want to reference back to the various media objects that we have encountered so far. These media objects are placed in several contexts, one of which is the right one corresponding to our path. However, we may also have different references, which lead to different contexts. So, the contexts we asked for in the second questions are just the contexts for the media objects.

As to the third question, we definitely want to have more information than just a name. Fortunately, the theory of media types is already based on the assumption of an underlying type system. Thus, we simply have to replace the names by values of any type allowed by the type system. Having defined such extended contexts, the query macros such as **look-up** and **cross-ref** would allow to traverse back a path in the story board and to explore alternative access paths, respectively.

However, one important aspect of media types is the use of classification abstraction. Conceptually, we do not define a set of media objects, but we generate them via queries defined on some underlying database schema. Therefore, we also need a conceptual abstraction for contexts.

In order to obtain this conceptual abstraction, we assume another base type *Context*, the values of which are context names. Instead of this, we could take the type URL, but in order to avoid confusion we use a new type.

A *context* consists of a name C, i.e. a value of type content, a type t_C and a defining query q_C, which must be defined on the media schema, i.e., the set of media types, such that

$$(\{(\text{object} : \mathit{URL}, \text{value} : t_C, \text{reference} : \mathit{Context})\}, q_C)$$

defines a view. Thus, executing the query q_C will result in a set of triples (u_i, v_i, r_i), where u_i is the URL of a media type, v_i is a value of type t_C, and r_i is the name of a context. If this context is undefined, this is interpreted as no reference for this object in this context. Note that in particular this definition of context leads to views over views.

4.4 Supertyping as a Special Case of Context Modelling

Let us finally reconsider the "old" definition of media types, which includes supertypes. In this case, all the supertypes are media types, thus depend on defining queries. They could be treated as queries defining a context. Thus, a media object of type M would be in as many contexts as there are supertypes of M. However, there are two important differences:

- In contextual information bases we want to select one context to obtain the information about the path, whereas the supertyping assumes that the combination of all supertypes defines the required context.
- If the supertypes are treated as if they are defining contexts, then there will be no references from their objects to other contexts. This omits the possibility of navigating through contexts.

Alternatively, we could take all the defining queries of supertypes of M together to define a context. Then each media object would belong to exactly one context, and as before there would not be any references to other contexts. Thus, supertyping turns out to be a simplified, static version of context modelling.

5 Conclusion

In this paper we investigated the question, how contexts of web information systems can be modelled in order to avoid customers getting lost or confused while navigating through such a system. We presented an approach based on the notion of context space. Depending on the location of a customer in the information space of the web information system, the usage history and a corresponding customer type a suitable continuation will be suggested to a customer, which will allow him or her to regain the possibility of using the respective WIS efficiently.

We then made a first step towards formally integrating context nodelling into the co-design methodology for WIS on the basis of the theory of contextual information bases developed by Teodorakis et al. In this theory a context is a collection of objects, each associated with a name and an optional reference to another object. Thus, both the name and the reference depend on the context. We slightly generalised this definition allowing now complex values to be used instead of only names. Furthermore, we considered media objects. With this generalisation we could show that the combination of the theory of media types with the theory of contexts is indeed possible.

The co-design methodology and thus also the theory of media types has already proven its usefulness by being applied in several very large scale development projects. We believe that the contextual extension developed in this article will further enhance the usefulness and applicability of the methodology. However, experiments need to be conducted that support our belief with empirical data.

References

1. M. Akaishi, N. Spyratos, Y. Tanaka. A Component-Based Application Framework for Context-Driven Information Access. In H. Kangassalo et al. (Eds.). *Information Modelling and Knowledge Bases* XIII: 254–265. IOS Press 2002.
2. P. Atzeni, A. Gupta, S. Sarawagi. Design and Maintenance of Data-Intensive Web-Sites. *Proc. EDBT '98*: 436–450. Springer LNCS 1377, Berlin 1998.
3. L. Baresi, F. Garzotto, P. Paolini. From web sites to web applications: New issues for conceptual modeling. *ER Workshops 2000*: 89–100. Springer LNCS 1921, Berlin 2000.
4. A. Bonifati, S. Ceri, P. Fraternali, A. Maurino. Building multi-device, content-centric applications using WebML and the W3I3 tool suite. *ER Workshops 2000*: 64–75. Springer LNCS 1921, Berlin 2000.
5. P. Bouquet, L. Serafini, P. Brezillon, M. Benerecetti, F. Castellani (Eds.). *Modeling and Using Context, CONTEXT'99 Proceedings*. Springer LNAI 1688, Berlin 1999.
6. A. Düsterhöft, B. Thalheim. *SiteLang: Conceptual Modeling of Internet Sites*. In H.S. Kunii et al. (Eds.). *Conceptual Modeling – ER 2001*: 179–192. Springer LNCS vol. 2224, Berlin 2001.
7. T. Feyer, K.-D. Schewe, B. Thalheim. Conceptual Modelling and Development of Information Services. In T.W. Ling, S. Ram (Eds.). *Conceptual Modeling – ER '98*: 7–20. Springer LNCS 1507, Berlin 1998.
8. R. Hirschheim, H. K. Klein, K. Lyytinen. *Information Systems Development and Data Modeling, Conceptual and Philosophical Foundations*. Cambridge University Press, Cambridge 1995.
9. T. Feyer, O. Kao, K.-D. Schewe, B. Thalheim. Design of Data-Intensive Web-Based Information Services. In *Proc. 1st International Conference on Web Information Systems Engineering*. Hong Kong (China). IEEE 2000.
10. F. Garzotto, P. Paolini, D. Schwabe. HDM - A model-based approach to hypertext application design. *ACM ToIS* vol. 11(1): 1–26, 1993.
11. M. Kirchberg, K.-D. Schewe, A. Tretiakov. Using XML to Support Media Types. Massey University 2002. submitted for publication.
12. H.C. Mayr, P.C. Lockemann, M. Bever. A Framework for Application Systems Engineering. *Information Systems* vol. 10(1): 97–111, 1985.
13. J. Mylopoulos, R. Motschnig-Pitrik. Partitioning Information Bases with Contexts. *Proc. CoopIS '95*: 44–55.

14. G. Rossi, D. Schwabe, F. Lyardet. Web Application Models are more than Conceptual Models. In P.P. Chen et al. (Eds.). *Advances in Conceptual Modeling*: 239–252. Springer LNCS 1727, Berlin 1999.
15. K.-D. Schewe, B. Thalheim. Modeling Interaction and Media Objects. In E. Métais (Ed.). *Proc. 5th Int. Conf. on Applications of Natural Language to Information Systems*: 313–324. Springer LNCS 1959, Berlin 2001.
16. K.-D. Schewe, B. Thalheim. *Conceptual Modelling of Internet Sites*. Tutorial Notes. ER'2000.
17. D. Schwabe, G. Rossi. An Object Oriented Approach to Web-Based Application Design. *TAPOS* vol. 4(4): 207–225. 1998.
18. M. Teodorakis, A. Analyti, P. Constantopoulos, N. Spyratos. Context in Information Bases. *Proc. CoopIS '98*: 260–270.
19. M. Teodorakis, A. Analyti, P. Constantopoulos, N. Spyratos. Querying Contextualized Information Bases. *Proc. ICT & P '99*. Plovdiv (Bulgaria) 1999.
20. B. Thalheim. Development of database-backed information services for Cottbus-Net. Report CS-20-97, BTU Cottbus 1997.
21. C. Wallace, C. Matthews. Communication: Key to Success on the Web. *Proc. eCoMo 2002*. Springer LNCS.

Using Message-oriented Middleware for Reliable Web Services Messaging

Stefan Tai, Thomas A. Mikalsen, and Isabelle Rouvellou

IBM T.J. Watson Research Center, Hawthorne, New York, USA
{stai,tommi,rouvellou}@us.ibm.com

Abstract. Web Services hold the promise of a standards-based platform for automating the integration of applications over diverse networks, operating systems and programming languages. Reliable messaging is critical in this context; many enterprise systems require a messaging infrastructure that guarantees message delivery even in the presence of software and network failures. Using existing message-oriented middleware (MOM) for reliable Web services messaging seems natural. However, a variety of implementation challenges, including the support for specific reliable Web services messaging protocols, must be addressed. In this paper, we discuss the options for and implications of employing MOM to implement reliable messaging for Web services. In doing so, we contribute to the understanding of reliability for Web services in general.

1 Introduction

Web services are applications that are described, published, and accessed over the Web using open XML standards. They promote a service-oriented computing model where an application exposes, using the Web Services Description Language (WSDL) [8] both its functionality (in a platform independent fashion) as well as its mappings to available communication protocols and deployed service implementations. This description can be discovered by client applications using service registries in order for the client to then use the service by means of XML messaging.

A key goal of the Web services framework is to provide a standards-based abstraction layer over diverse network transports, operating systems, and programming languages, and therefore, to provide a platform for automating the integration of these diverse systems and their applications. Further, the framework aims to support application/system integration both within and between organizational boundaries. As Web services reach their full potential, new services will emerge that are based on compositions of other Web services and specified using interoperable business process standards.

If this vision is to be realized, basic service interactions, spanning technological and organizational boundaries, must be reliable. Consequently, new protocols for reliable messaging for Web services have been proposed [3] [9] [22] [15]. Yet, the use of existing message-oriented middleware (MOM) for reliable Web services messaging seems reasonable, too. There are now many different options for implementing reliable messaging for Web services, with various implications on overall application-to-application reliability.

C. Bussler et al. (Eds.): WES 2003, LNCS 3095, pp. 89–104, 2004.

This paper provides some clarity about reliable messaging and using message-oriented middleware for Web services. We first discuss different aspects of reliable messaging, identifying the facets of MOM that affect reliability. We then explore different, exemplary options for applying existing messaging technologies to achieve reliable messaging for Web services. We explain the options in detail, and assess them with respect to the level of reliability they achieve and the assumptions they place on the implementation of the distributed services.

2   Messaging Middleware Reliability

In distributed systems, multiple processes (executing on different nodes in a network) interact by sending and receiving messages. In this environment, it is often desirable to decouple the sending and receiving processes such that forward progress can be made in the presence of failures and temporary unavailability. To achieve this loose-coupling of processes, message-oriented middleware is often employed.

Messaging middleware is specialized software that accepts messages from sending processes and delivers them to receiving processes (typically across a network). Such middleware typically support two common delivery patterns: point-to-point (p2p) and publish/subscribe (pub/sub). The architecture of the middleware can take many forms. The two principle styles are centralized and distributed architectures. With centralized architectures, all processes communicate with a common messaging server. In distributed architectures, processes communicate with local messaging middleware components; these local messaging components then communicate over the network to deliver messages on behalf of senders and receivers. In this paper, we will focus on distributed architectures, as these are natural in a Web environment.

2.1   Aspects of Reliability

Messaging middleware can perform various functions that facilitate reliable, loose-coupling. In distributed architectures, the sender's messaging component can tolerate network failures by repeatedly sending a message until it is acknowledged by the receiver's messaging component; this interaction can occur even after the sending process has terminated (or is otherwise unavailable). The receiver's messaging component can tolerate the unavailability of the receiving process by maintaining messages until the receiving process is ready. From the sender application's viewpoint, this allows the application to "fire-and-forget" messages, relying on the middleware to guarantee message delivery.

The messaging endpoints ideally are the messaging clients (the sending and receiving applications). With common messaging middleware, the guarantees are, however, typically restricted to the middleware endpoints of message brokers and message queue managers. It is then assumed that a messaging client accesses a messaging middleware endpoint locally and using transactions (though a distributed and/or non-transactional access is possible, too). The middleware only ensures message delivery within its own "network" of managerial messaging endpoints.

In addition to acknowledged delivery (through proper correlation of messages and acknowledgments), ordered delivery of messages is another aspect of reliable messaging middleware. This is particularly important in asynchronous environments, where messages are typically stored by the middleware (for example, a message queue) before they are dispatched to or retrieved by final recipients.

Messages can further be attributed with expiration timestamps, a priority attribute, a reply-to address, and other properties that contribute to messaging reliability. Such attributes are checked by the middleware in order to prevent delivery of a message if it is no longer valid, to prioritize messages that are stored at the middleware for later dispatch, and to guarantee that a receipt acknowledgment is sent back to a specific, application-defined reply address.

A further important aspect of reliability concerns the integration of a message delivery in a larger processing context. A message typically is part of some business process (messaging conversation) and atomic unit-of-work. Key requirements on reliable messaging middleware therefore include the ability to atomically group a message with other messages and other process activities, and, to integrate a message store like a message queue as a resource manager in a distribution transaction.

2.2 Three Facets of Reliability

The above described aspects of reliability lead to the definition of three general facets of reliability:

Middleware endpoint-to-endpoint reliability
A message, once delivered from an application (process) to the messaging middleware, is guaranteed to be (eventually) available for consumption by the receiving process. The middleware ensures eventual message delivery within its distributed network of middleware endpoints.

Application-to-middleware reliability
The middleware's messaging API, used to send and receive messages, supports reliability properties such as message delivery guarantees, message persistence, and transactional messaging.

Application-to-application reliability
Sending and receiving applications engage in transactional business processes that rely on application-to-middleware reliability and middleware endpoint-to-endpoint reliability.

2.3 Technologies

Each facet of reliable messaging imposes certain requirements on the middleware. Fundamental requirements include the integration of a persistent message store and the implementation of a reliable protocol to move messages between these persistent stores.

Examples of message-oriented middleware products implementing reliable messaging are IBM's Websphere MQ (formerly called MQSeries) [10], TIBCO's Rendezvous [21], and Microsoft's MSMQ [12] middleware. They each support their own proprietary messaging APIs and protocols, as well as the standard Java messaging API, the Java Message Service (JMS) [16].

While these products provide a complete and proprietary solution to reliable messaging, a number of open standards for reliable messaging have been emerging. These include the ebXML Message Service [15], the HTTPR protocol [22], the WS-Reliability protocol [9], and the WS-ReliableMessaging protocol [3]. The objective of these specifications is not to promote or prescribe a particular product and middleware infrastructure, but to define the rules governing message delivery, such as for message correlation, persistence, and acknowledgments. HTTPR, for example, can be supported by different messaging agent implementations that are paired with some persistent store; reliable messaging thereby is introduced over the (unreliable) HTTP protocol. WS-Reliability and WS-ReliableMessaging define an XML messaging protocol that allows reliability to be introduced to SOAP messages independent of the underlying transports.

3   Web Services Messaging Using SOAP

The messaging protocol most commonly used for Web services is the Simple Object Access Protocol (SOAP) [5]. SOAP is an XML message format (of an envelope, message headers, and a message body) and a standard encoding mechanism that allows messages to be sent over a variety of transports, including HTTP and JMS.

SOAP-over-HTTP has been the most popular choice for Web services messaging, as it is a well-understood messaging model that is easy to implement and maintain. Due to the frequent use of SOAP-over-HTTP, SOAP is often understood to be a request-response (RPC-like) protocol. However, the synchronous flavor of SOAP-over-HTTP results more from HTTP than from SOAP. If JMS is used as a transport, for example, SOAP can be used to implement asynchronous messaging.

SOAP messaging can take different forms of reliability depending on the underlying transport chosen. While SOAP-over-HTTP is not reliable, SOAP-over-HTTPR ensures that messages are delivered to their specified destination. Similar reliability guarantees can be made for SOAP-over-JMS and SOAP-over-Websphere MQ. On the other hand, a SOAP message itself can be extended to include reliability properties, using the recently proposed WS-Reliability or WS-ReliableMessaging standards. These "extended SOAP" messages then carry relevant reliability information that must be understood and supported by a messaging infrastructure (that may or may not employ other reliable messaging technology such as Websphere MQ).

It is important to note that SOAP is both a message format and a transport-flexible messaging protocol. With SOAP messaging, we consequently refer to the use of SOAP as both a format and a protocol. To clarify the use of SOAP as a protocol, we refer "SOAP-over-<X>", where <X> is the chosen transport.

SOAP Messaging implies the use of a SOAP library (such as Apache AXIS [1]) to construct, send, receive, and parse SOAP messages, using the desired transport and encoding rules; an application involved in SOAP messaging interfaces the SOAP library, but does not interface any deployed middleware that implements the SOAP transport. If, on the other hand, a middleware like Websphere MQ is used as the messaging protocol and middleware, and SOAP-formatted XML messages are being exchanged (perhaps with the help of some utilities for manipulating SOAP enve-lopes), we refer to MQ messaging using SOAP/XML messages.

Presently, SOAP as a messaging protocol is not particularly feature rich, making for subtle distinction between SOAP-over-JMS and JMS messaging using SOAP messages (for example.) However, this will change as SOAP matures; the proposed standards for reliable messaging and other specifications addressing, for example, message routing and referral [13] [14] can be seen as evidence for this.

4 Reliable Messaging for Web Services

Reliable messaging for Web services is about achieving reliable messaging aspects (as described in Section 2) within the Web services environment. In this section, we study how existing reliable messaging technologies can be used for this purpose.

A number of reliable messaging technologies exist. These include enterprise mes-sage-oriented middleware like IBM Websphere MQ, distributed object messaging standards like JMS, or reliable transport protocols for Web environments like HTTPR. An application may choose to use either one of these technologies, or a combination of these technologies, to address reliable Web services messaging. An application may also, in addition or as an alternative to the above options, choose to implement reliability mechanisms itself as part of the application. Consequently, reli-ability may (or may not) be addressed on any or all of the protocol/transport, middle-ware, and application layers.

Figure 1 illustrates the resulting set of common options for implementing reliable messaging in a Web services world:

a) SOAP (with or without a reliability protocol like WS-ReliableMessaging) is used with an unreliable transport (like HTTP); reliability mechanisms are implemented on the application/SOAP messaging layer

b) A reliable transport protocol like HTTPR is used for SOAP messaging (without a reliability protocol like WS-ReliableMessaging); a middleware system (that is, any implementation of messaging agents supporting HTTPR based on HTTP and some persistent storage capability) is required

c) A reliable, proprietary middleware system like IBM Webpshere MQ is used for SOAP messaging (without a reliability protocol like WS-ReliableMessaging); the middleware defines the transport protocol and provides the necessary distributed infrastructure

d) A reliable messaging standard like JMS is used for SOAP messaging (without a reliability protocol like WS-ReliableMessaging); a JMS implementation is re-quired

e) A reliable, proprietary middleware system like IBM Webpshere MQ (any middleware system leveraged for Web services and with some means of durable storage for message logging qualifies) is directly used, independent of SOAP

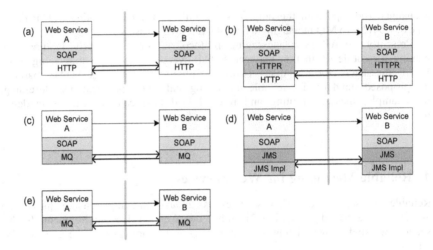

Fig. 1. Reliable Messaging Implementation Options

5 Assessment

As defined in Section 2.2, three facets of reliability exist:

- middleware endpoint-to-endpoint reliability
- application-to-middleware (and middleware-to-application) reliability, and
- application-to-application reliability

In the following, we first examine the above listed options with respect to middleware endpoint-to-endpoint reliability and application-to-middleware reliability. We identify which (required or additional) responsibilities the application developer has when choosing an option. These discussions set the stage to discuss application-to-application reliability.

5.1 Middleware Endpoint-to-Endpoint Reliability

Middleware endpoint-to-endpoint reliability assumes that a middleware is being used for mediation between communicating messaging partners. Middleware endpoint mediation essentially refers to the idea that messages are made persistent locally on the sender and receiver sides before and after they are being sent (Figure 2). All messages are given unique identifiers, so that a message sender (the endpoint on the sender side) can re-send a message until it gets a positive acknowledgment of receipt by the receiver (the endpoint on the receiver side).

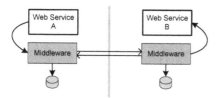

Fig. 2. Middleware Mediation

Option (a). Basic SOAP-over-HTTP does not support endpoint-to-endpoint reliability. HTTP is a synchronous protocol that is "reliable" as long as the connection stays alive (like TCP): it delivers messages at most once, in order, and with a definite acknowledgment for each message delivery or delivery failure. HTTP is unreliable in the sense that when a connection is lost, the message sender will get a connection failure event, but be in doubt about the status of the message:

• The message might not have been delivered to the destination.
• The message might have been delivered to the destination and might have been processed by the destination; the receiver might have replied, or, the receiver might know about the connection failure and may or may not have attempted to rollback its processing, if possible.

To address this uncertainty, the application can mimic middleware mediation on the application layer: The message sender application keeps persistent copies of messages before sending them and attempts message delivery until positive, explicit application acknowledgments confirm the receipt. The receiver application implements the equivalent functionality on its side for reliably sending replies.

If a reliability protocol like WS-Reliability or WS-ReliableMessaging is used, reliability information (for message tracking, for example) is encoded in the SOAP headers. The use of these "standards" allows to ensure endpoint-to-endpoint reliability across distributed services, independent of how the applications choose to implement the reliability mechanisms.

Encoding (lower-level) reliability mechanisms with SOAP messages may however confuse the implementation of higher-level business processes (depending on how the SOAP messages with reliability extensions are created). Furthermore, robust reliable messaging middleware already exist; in cases where a reliable messaging transport for SOAP is used, supporting reliability on the SOAP and on underlying transport layers may cause avoidable inefficiency.

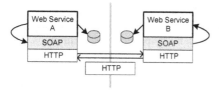

Fig. 3. SOAP-over-HTTP

Option (b). SOAP-over-HTTPR supports middleware endpoint-to-endpoint reliability. The middleware endpoints are the (implementations of) HTTPR messaging agents combined with some persistent storage capability.

An application using this option can either choose to implement the HTTPR messaging agents itself, or select an existing available implementation (such as [11]); additionally, a persistent store at the sender and receiver ends must be installed.

An application sends a SOAP message using a SOAP library (for example, AXIS) that is configured to use an HTTPR sender agent. The SOAP message itself is not HTTPR-specific; the message has the same format as a SOAP-over-HTTP message. The SOAP message is then sent as the body of an HTTPR request; additional required information of message id and correlation information are carried in the HTTPR message context header preceding the HTTPR body. On the receiver side, an HTTPR agent receives the message and directs it further to the SOAP layer. The application's WSDL definitions reflect the use of HTTPR in the bindings specification.

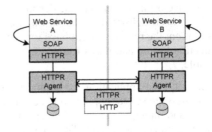

Fig. 4. SOAP-over-HTTPR

Option (c). SOAP-over-MQ also supports middleware endpoint-to-endpoint reliability. The middleware endpoints are message queue managers provided by the messaging middleware product; the persistent stores are message queues.

IBM Websphere MQ is one example of a proprietary message queuing system that supports SOAP messaging. A typical scenario will require the sender and receiver applications to each set up a queue manager (which may be part of a single MQ cluster) and to define a service queue on the receiver side and a response queue on the sender side. The sending application puts a message into the receiver's service queue, and the receiving application consumes the message from the service queue and posts a reply to the sender's queue. A SOAP client library is provided for an application to send SOAP messages over Websphere MQ. On the server side, a SOAP handler processes the incoming messages and invokes the appropriate deployed application.

Note that the message delivery pattern is asynchronous. That is, different from the synchronous HTTP request-response style, MQ (and JMS, see below) request-reply messaging is implemented as two unidirectional, correlated messages (request message and reply message). The sending application is not blocked waiting for the reply, tolerating non-availability of the receiving application.

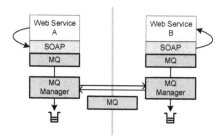

Fig. 5. SOAP-over-MQ

Option (d). SOAP-over-JMS principally compares to SOAP-over-HTTPR and SOAP-over-MQ. Similar to the HTTPR option, a middleware system implementing the JMS standard is required. Any (single) JMS implementation can be used; in case of IBM Websphere MQ, the reliability for SOAP-over-JMS messaging is basically the same as for SOAP-over-MQ.

The middleware endpoints are JMS senders and JMS receivers. Attention must be paid to the different receiver kinds possible: a simple Java client using a JMS MessageListener is not as robust and transactionally reliable as an Enterprise JavaBean (EJB) or a Message-Driven Bean (MDB) (see also Section 5.2).

Using JMS as a transport for SOAP messaging is vendor-proprietary, as (different from SOAP-over-HTTP) no standard has been defined. JMS implementations can also significantly differ from each other (for example, the JMS wire protocol is also vendor-proprietary), so that the SOAP client and the Web services provider must use the same runtime environment to achieve endpoint-to-endpoint reliability (that is, the same vendors' JMS implementation and SOAP-to-JMS transport binding library are required).

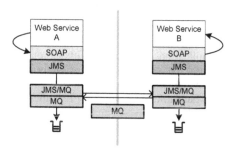

Fig. 6. SOAP-over-JMS

Option (e). Middleware endpoint-to-endpoint reliability can also be achieved by employing enterprise messaging middleware directly, independent of SOAP.

In this case, the middleware must be leveraged to support XML messaging. For example, the middleware may provide an adapter component that takes an XML message (including a standard SOAP message) from a sending application, and then packages the message into its own distributed messaging formats and protocols (such as MQ messages, or JMS messages). The adapter then reliably sends the message to a defined remote queue, using the distributed messaging middleware. On the receiving side, another middleware adapter component reads the message from the queue, unpacks the XML message contained in it, and sends the message to the Web service that is to be invoked.

Benefits of this option include the use of a robust distributed messaging network that may already exist. Furthermore, few changes are required for the sending and receiving application; the invoked Web service may not need to be modified at all.

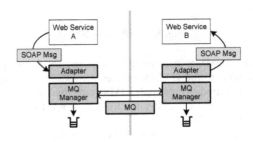

Fig. 7. MQ

5.2 Application-to-Middleware Reliability

Application-to-middleware reliability refers to the reliability features provided by the middleware's application-to-endpoint interface (a.k.a., the messaging API). These may include the following:

- message delivery guarantees (e.g., exactly-once, at-most-once, at-least-once);
- fault-tolerant invocation (of the messaging endpoint);
- the ability to atomically group messaging operations with other application actions.

Depending on the architecture of the messaging middleware, supporting such features poses different challenges. For example, in distributed architectures, where the endpoint is local to the application, fault-tolerant invocation may not be an issue and exactly-once delivery might be supported by the endpoint-to-endpoint reliability mechanisms described above. However, in centralized architectures, where there is a shared messaging server, fault-tolerant invocation becomes more difficult, whereas delivery guarantees become less so. (Hybrid architectures, with centralized messaging servers that are themselves distributed exhibit additional challenges.)

The ability to atomically group messaging operations (e.g., sending a message) with other application activities (e.g., updating a database) may require that the messaging endpoint act as a participant in a two-phase commit protocol (2PC) [2]. For example, if the endpoint is a remote Web service, the WS-Coordination framework [6] and the WS-Transaction "Atomic Transaction" protocols [7] could be used for this purpose; if the endpoint is local, the Java Transaction API [17] or J2EE Connector Architecture [18] could be used.

When assessing the options above with respect to application-to-middleware reliability, we must observe that SOAP itself does not define any such reliability features.

Therefore, a SOAP library that supports application-to-middleware reliability features, such as transactionally coordinating messaging operations with other application activities, may have to define proprietary SOAP APIs.

Option (a). When using SOAP-over-HTTP (with or without a reliability protocol for SOAP), the reliability mechanisms may be implemented as part of the application. The application can therefore ensure that its components for creating, storing, and delivering messages are all accessed reliably. For example, to support the atomic grouping of message creation with other application activities, the message store might be implemented as a resource manager that is transactionally coordinated with other resource managers used by the application. Figure 8 illustrates the use of local transactions for this purpose.

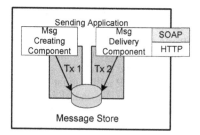

Fig. 8. Local Transactions for Message Storage

In the figure, the transaction that puts a message in the message store (Tx1) typically will further comprise other application activities, thereby grouping the message storage with other activities in an atomic unit-of-work. A second transaction (Tx2) can be defined to read (copy) messages from the message store and to initiate the remote message delivery.

Option (b) to (d). In the cases of SOAP messaging over a reliable transport and middleware (SOAP-over-HTTPR, SOAP-over-MQ, and SOAP-over-JMS/MQ), the underlying middleware implements the reliability mechanisms for endpoint-to-endpoint reliability (including persistent message storage). The application-to-middleware guarantees that can be provided are the message delivery guarantees of the underlying transport and, if the endpoint is local, the reliability of the local procedure calls between the application and the local endpoint. However, unlike Option (a), the application cannot transactionally group a messaging operation with other activities of a larger business process.

Figure 9 illustrates this case. The (sending or receiving) application only interfaces the SOAP library and does not use the underlying middleware directly. The application uses local calls for SOAP messaging, independent of any functionality that the underlying reliable middleware may offer.

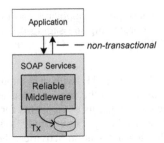

Fig. 9. Non-transactional Use of SOAP Library

If the application requires additional reliability features, in particular the ability to transactionally group messaging operations with other activities as described above, these options are not sufficient. Either the SOAP library would have to expose underlying reliability features (using proprietary extensions) or the application would have to access the underlying middleware directly when these features are needed. However, such solutions may have negative consequences: proprietary extensions will not likely be compatible with future SOAP API standards that do address application-to-middleware reliability, while accessing the underlying middleware directly could interfere with the SOAP libraries use of the middleware.

Option (e). The "direct middleware" option of using JMS or MQ for XML messaging describes the case where the application uses the reliable middleware's API to send and receive messages. Therefore, application-to-middleware reliability relates to the direct use of the underlying middleware's API and its reliability features.

JMS and MQ support, for example, the notion of a "transacted session", allowing a message sender to group a number of messages into an atomic unit-of-work. The middleware then ensures that either all messages of the group are delivered, or none of them. With the use of persistent message queues, support for resource management as required by distributed object transactions is also provided. Therefore, messages put to a message queue can be transactionally coupled with other database and distributed invocations. Figure 10 illustrates this transactional feature.

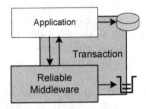

Fig. 10. Transactional Use of Reliable Middleware

5.3 Application-to-Application Reliability

Middleware endpoint-to-endpoint reliability and application-to-middleware reliability provide the foundation on which (higher-level) distributed business processes can be developed. These distributed business processes are constructed from basic application-to-application interactions. Reliability of these basic application-to-application interactions is therefore critical to the reliability of the business process as a whole.

In the following, we briefly discuss ways to support basic application-to-application reliability by traditional means of direct transaction and queued transaction processing [2].

5.3.1 Direct Transaction Processing

With direct transaction processing, an agreement protocol (e.g., two-phase commit) is used to directly include one application's transaction processing as part of another application's transaction processing. This is shown in Figure 11.

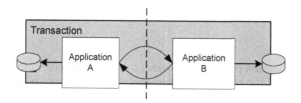

Fig. 11. Direct Transaction Processing

In the figure, Application A and application B interact within the same global transaction. If there is a failure, or if either application decides that the outcome or effect of the interaction is inconsistent, the transaction can be aborted, returning both applications to a consistent state. Furthermore, direct transaction processing can include more than two applications, extending the scope of the reliability guarantee.

The WS-Coordination and WS-Transaction specifications define an interoperable mechanism for supporting direct transaction processing over the Web.

5.3.2 Queued Transaction Processing

With queued transaction processing, applications interact indirectly using reliable message-oriented middleware; a transaction service, that can integrate messaging resources (e.g., queues) with other resources used by the applications (e.g., databases), is also needed. Employing queued transaction processing for reliable request-reply interactions is shown in Figure 12.

In the figure, three separate transactions are used to ensure that this basic application-to-application interaction is reliable. In transaction 1, application A access local resources and sends a request message to application B; however, the message is not visible to application B until transaction 1 commits. In transaction 2, application B consumes the message, accesses local resources, and sends a response message to application A; the response is not visible to application A until transaction 2 commits. In transaction 3, application A consumes the response message from application B and accesses local resources. A failure during any single transaction returns the interaction to a well defined, and locally consistent, state.

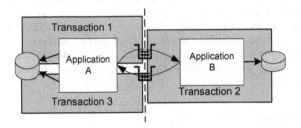

Fig. 12. Queued Transaction Processing

Reliable messaging middleware, supporting application-to-middleware and endpoint-to-endpoint reliability, as described above, can be employed to support queued transaction processing on the web.

5.3.3 Discussion

While direct transaction processing provides a reliable mechanism for multiple applications to agree on the effect and outcome of some joint processing, it can impose tight application coupling, reducing the autonomy the applications involved: for example, 2PC protocols typically require that all participants be active at the same time and can force participants to hold locks (on their local resources) on behalf of other participants.

Yet, queued transaction processing, which supports a looser coupling of applications, can force applications to implement potentially complex logic for coordinating the effects and outcomes of message delivery and processing with the effects and outcomes of other activities (e.g., database updates); if, for example, the response message in Figure 12 indicates that the outcome of application B's processing is inconsistent with application A's processing, application A cannot simply abort its transaction and expect to return to a consistent state (as was the case with direct transaction processing); rather, application A will have to actively correct for the inconsistency, perhaps initiating additional transactions and interactions with application B (for example, a compensating transaction that reverses the work performed in transaction 1).

Our work on conditional messaging [19] and Dependency Spheres [20] propose additional techniques for application-to-application reliability that overcome some of these disadvantages.

6   Summary

In this paper, we presented options for implementing reliable messaging for Web services using existing messaging technology. For each option, we discussed how middleware endpoint-to-endpoint and application-to-middleware reliability is achieved. We further showed how reliable messaging, when combined with traditional transaction processing techniques (such as direct transaction processing and queued transaction processing), can be applied to support basic application-to-application reliability.

We focused our discussion on the middleware infrastructure for implementing reliable messaging for Web services, as the middleware concern will always be a key challenge independent of which (existing or emerging, future standard) reliable messaging protocol will be used (or if no standard is used). In fact, we believe that any proposed reliable messaging standard must be assessed in consideration of supporting middleware implementations. As we have shown in this paper, there are subtle, but important differences of middleware implementations of Web services that determine to what extent two deployed Web services can exchange messages reliably. This observation may help to estimate implementation efforts and costs related to reliable messaging and therefore, to choose one of the different possible (standards- or non-standards-based) options.

References

1. Apache AXIS. http://xml.apache.org/axis/
2. P. A. Bernstein, E. Newcomer. Principles of transaction processing. Morgan Kaufmann, 1997.
3. R. Bilorusets et al. Web Services Reliable Messaging Protocol (WS-ReliableMessaging). BEA, IBM, Microsoft, March 2003. ftp://www6.software.ibm.com/software/developer/library/ws-reliablemessaging.pdf
4. A. Bosworth et al. Web Services Addressing (WS-Adressing). BEA, IBM, Microsoft, March 2003. ftp://www6.software.ibm.com/software/developer/library/ws-addressing.pdf
5. D. Box et al. Simple Object Access Protocol (SOAP) 1.1. W3C Note 08 May 2000. http://www.w3.org/TR/2000/NOTE-SOAP-20000508/
6. F. Cabrera et al. Web Services Coordination (WS-Coordination). BEA, IBM, Microsoft, August 2002. http://www-106.ibm.com/developerworks/library/ws-coor/
7. F. Cabrera et al. Web Services Transaction (WS-Transaction). BEA, IBM, Microsoft, August 2002. http://www-106.ibm.com/developerworks/library/ws-transpec/
8. E. Christensen et al. Web Services Description Language (WSDL) 1.1. W3C Note 15 March 2001. http://www.w3.org/TR/wsdl
9. C. Evans et al. Web Services Reliability (WS-Reliability), Version 1.0. Fujitsu, Hitachi, NEC, Oracle, Sonic Software, Sun Microsystems 2003. http://xml.fujitsu.com/en/about/WS-ReliabilityV1.0.pdf
10. IBM Corp. IBM WebSphere MQ. http://www-3.ibm.com/software/ts/mqseries/messaging/
11. IBM Corp. WebSphere MQ Support for Web Services and HTTPR. MA0R Support Pac, IBM Corporation, April 2002. http://www-3.ibm.com/software/ts/mqseries/txppacs/ma0r.html
12. Microsoft. Microsoft Message Queuing (MSMQ). http://www.microsoft.com/msmq/default.htm
13. H. F. Nielsen, S. Thatte. Web Services Routing Protocol (WS-Routing), Microsoft, October 2001. http://msdn.microsoft.com/library/en-us/dnglobspec/html/ws-routing.asp
14. H. F. Nielsen et al. Web Services Referral Protocol (WS-Referral), Microsoft, October 2001. http://msdn.microsoft.com/library/en-us/dnglobspec/html/ws-referral.asp
15. OASIS. ebXML Message Service Specification Version 2.0. OASIS, April 2002. http://www.oasis-open.org/committees/ebxml-msg/documents/ebMS_v2_0.pdf
16. Sun Microsystems. Java Message Service API Specification v1.1. Sun Microsystems, April 2002. http://java.sun.com/products/jms/
17. Sun Microsystems. Java Transaction API (JTA), Version 1.0.1B. Sun Microsystems, November 2002. http://java.sun.com/products/jta/

18. Sun Microsystems. Java 2 Enterprise Edition: J2EE Connector Architecture Specification, Version 1.0. Sun Microsystems, August 2001 http://java.sun.com/j2ee/connector/
19. S. Tai, T. Mikalsen, I. Rouvellou, S. Sutton. Conditional Messaging: Extending Reliable Messaging with Application Conditions. Proceedings of the 22nd IEEE International Conference on Distributed Computing Systems (ICDCS 2002, Vienna, Austria), IEEE, pp. 123–132, July 2002
20. S. Tai, T. Mikalsen, I. Rouvellou, S. Sutton. Dependency-Spheres: A Global Transaction Context for Distributed Objects and Messages. Proceedings of the 5th IEEE International Enterprise Distributed Object Computing Conference (EDOC 2001, Seattle, USA), IEEE, pp. 105–115, September 2001
21. TIBCO. TIBCO Rendezvous. http://www.tibco.com/solutions/products/active_enterprise/ rv/default.jsp
22. S. Todd, F. Parr, M. Conner. A Primer for HTTPR. An Overview of the Reliable HTTP Protocol. IBM Corporation, July 2001. http://www-106.ibm.com/developerworks/ webservices/library/ws-phtt/

Reusability Constructs in the Web Service Offerings Language (WSOL)

Vladimir Tosic, Kruti Patel, and Bernard Pagurek

Department of Systems and Computer Engineering, Carleton University,
1125 Colonel By Drive, Ottawa, Ontario, K1S 5B6, Canada
{vladimir, bernie}@sce.carleton.ca, kruts.patel@lycos.com

Abstract. The Web Service Offerings Language (WSOL) is a novel language for the formal specification of classes of service, various types of constraint, and management statements for Web Services. Compared with recent competing works, WSOL has several unique characteristics. One of them is a diverse set of reusability constructs: service offerings, constraint groups, constraint group templates, extension, inclusion, applicability domains, and operation calls. These constructs enable sharing parts of WSOL specifications between classes of service of different Web Services and development of libraries of reusable WSOL specifications. Consequently, they can help in alleviating heterogeneity of Web Services. In addition, reusability constructs are useful for easier development of new WSOL specifications from existing ones, for easier selection of Web Services and their classes of service, and for dynamic (run-time) adaptation of relationships between provider and consumer Web Services. Integration of WSOL reusability constructs into the works competing with WSOL would be beneficial.

1 Introduction and Motivation

The Web Service Description Language (WSDL) version 1.1 is the de-facto standard for the description of Web Services. However, WSDL does not enable the specification of constraints, management statements, and classes of service for Web Service. As discussed in [1], the specification of different types of constraint and management statements is necessary for the management of Web Services and Web Service compositions. In addition, classes of service are a simple and lightweight alternative to contracts, service level agreements (SLAs), and profiles. Therefore, we have decided to develop our own language for the specification of classes of service, various types of constraint, and management statements for Web Service. We have named this language the **Web Service Offerings Language (WSOL)**.

When multiple classes of service are specified, there is often a lot of similar information that differs in some details. For example, classes of service that a telecommunication service provider offers its customers often have similarities. Analogously, two classes of service for the same Web Service can be the same in many elements, but differ only in response time and price. Defining common or similar parts of classes of service once and using these definitions many times simplifies the specification of new classes of service. Next, when it is explicitly stated that two classes of service share common parts, it is much easier to compare them. Such comparisons are

C. Bussler et al. (Eds.): WES 2003, LNCS 3095, pp. 105–119, 2004.
© Springer-Verlag Berlin Heidelberg 2004

useful in the process of selection and negotiation of Web Services and their classes of service. Further, when monitored classes of service have common elements, the overhead placed on the management infrastructure for the monitoring of Web Services, metering or calculation of quality of service (QoS) metrics, and evaluation of constraints, might be reduced. In addition, manipulation of classes of service can be used for simple dynamic (run-time) adaptation of Web Service compositions. Explicit specification of relationships between classes of service supports their comparison, as well as their manipulation.

For these reasons, we have built into WSOL a diverse set of **reusability constructs**. These constructs enable reuse of parts of WSOL specifications and easier comparisons of WSOL specifications, even when these WSOL specifications are specified for different Web Services. In this way, WSOL reusability constructs can help in alleviating heterogeneity of Web Services. They also model static relationships between classes of service (relationships that do not change during run-time) and thus support manipulation of classes of service.

WSOL was developed independently of and in parallel with several recent works that address issues somewhat similar to WSOL. However, these related works do not have such a diverse and rich set of reusability constructs. In our opinion, this is one of the advantages of WSOL [2]. We believe that integration of WSOL reusability constructs into the related works, as well as eventual future standards in this area, would be beneficial. Therefore, in this paper, we explain WSOL reusability constructs and their potential influences on related works. We assume that the reader is familiar with WSDL and the Extensible Markup Language (XML).

The paper is organized as follows. In this section, we have summarized the motivation for WSOL reusability constructs and the motivation for writing this paper. In the next section, we give a brief overview of WSOL and the most important related works. The core of the paper is Section 3, where we explain WSOL reusability constructs. Section 4 discusses potential influences between reusability constructs in WSOL and in major related works. In the final section, we summarize how the WSOL reusability constructs are useful for Web Services. The Appendix contains some WSOL examples of the discussed WSOL reusability constructs.

Our other publications on WSOL discuss and illustrate different aspects of WSOL and its management infrastructure, the Web Service Offerings Infrastructure (WSOI). [1] and [2] present WSOL and WSOI, [3] compares WSOL and related works, while [4] contains detailed information about the WSOL syntax and its examples. (Note that WSOL was improved since the publication of [4]). Our research report [5] is an extended version of this paper and contains examples and additional details on the topics we discuss hereafter.

2   A Brief Overview of WSOL and the Related Work

The Web Service Offerings Language (WSOL) is a language for the formal specification of classes of service, various types of constraint, and management statements for Web Services. The syntax of WSOL is defined using XML Schema, in a way compatible with WSDL 1.1. WSOL descriptions of Web Services are specified outside WSDL files.

The crucial concept in WSOL is a **service offering (SO)**. A WSOL service offering is the formal representation of a single class of service of one Web Service [1]. It can also be viewed as a simple contract or SLA between the provider Web Service, the consumer, and eventual management third parties. A Web Service can offer multiple service offerings to its consumers, but a consumer can use only one of them at a time in one session. A Web Service can have in parallel many open sessions, in some cases even several sessions with the same consumer. A WSOL service offering contains the formal definition of various types of constraint and management statements that characterize the represented class of service.

Every WSOL **constraint** contains a Boolean expression that states some condition (guarantee or requirement) to be evaluated. Boolean expressions in constraints can also contain arithmetic, date/time/duration, and some simple string expressions. The constraints can be evaluated before and/or after invocation of operations or periodically, at particular date/time instances. WSOL supports the formal specification of functional constraints (pre-, post-, and future-conditions), quality of service (QoS) constraints (describing performance, reliability, availability, and similar 'extra-functional' properties), and access rights (for differentiation of service).

A WSOL **statement** is any construct, other than a constraint, that states important management information about the represented class of service. WSOL enables the formal specification of statements about management responsibility, subscription prices, pay-per-use prices, and monetary penalties to be paid if constraints are not met. In addition, WSOL has extensibility mechanisms that enable definition of new types of constraint and management statement as XML Schemas.

Apart from definitions of constraints and management statements, service offerings can contain reusability constructs discussed in detail later in this paper. WSOL reusability constructs determine static relationships between service offerings. These relationships do not change during run-time. One example of a static relationship is when a service offering extends another service offering. In addition, WSOL enables specification of dynamic (i.e., run-time) relationships between service offerings (these relationships can change during run-time). Dynamic relationships are specified outside definitions of service offerings in the format presented in [1, 4]. A dynamic relationship states what class of service could be an appropriate replacement if a particular group of constraints from the used class of service cannot be met.

We use the term '**WSOL item**' to refer to a particular piece (service offering, constraint, statement, or reusability construct) of a WSOL specification. Some WSOL items can contain other WSOL items.

To verify the WSOL syntax, we have developed a WSOL parser called 'Premier' [4]. To enable monitoring of WSOL-enabled Web Services, metering and calculation of QoS metrics, evaluation of WSOL constraints, accounting and billing, as well as dynamic adaptation of compositions including WSOL-enabled Web Services, we are developing the Web Service Offerings Infrastructure [1, 2].

Our work on WSOL draws from the considerable previous work on the differentiation of classes of service in telecommunications and on the formal representation of various constraints in software engineering. In addition, our work on reusability constructs in WSOL is based upon many works on reusability constructs in other languages, both programming languages and description languages. In the next section, we explain how we have studied these existing reusability constructs for inclusion into WSOL.

In parallel with our work on WSOL, several XML languages with somewhat similar goals for Web Services have been developed. The IBM **Web Service Level**

Agreements (WSLA) [6] and the HP **Web Service Management Language (WSML)** [7] are two powerful languages for the formal XML-based specification of custom-made SLAs for Web Service. SLAs in these two languages contain only QoS constraints and management information. Another language that can be used for specification of SLAs for Web Services is **SLAng** [8]. SLAng enables specification of SLAs not only on the Web Service level, so it has broader scope than WSLA and WSML. However, the definitions of QoS metrics are built into SLAng schema, so SLAs have predefined format. Since the current version of SLAng lacks flexibility and power, it seems less well suited for Web Services than WSLA and WSML. Next, **WS-Policy** [9] is a general framework for the specification of policies for Web Services. A policy can describe any property of a Web Service or its parts, so it corresponds to WSOL concepts of a constraint and a statement. The detailed specification for particular categories of policies will be defined in specialized languages. Currently, specification details are defined only for security policies and to some extent for functional constraints, but not yet for QoS policies, prices/penalties, and other management issues. **DAML-S (DAML-Services)** [10] and the **Web Service Modeling Framework (WSMF)** [11] are examples of other works related to WSOL.

The main distinctive characteristics of WSOL, compared with the mentioned competing works, are: support for classes of service and their static and dynamic relationships, support for various types of constraint and statement, a diverse set of reusability constructs, features reducing run-time overhead, and support for management applications. A more detailed comparison between WSOL and some of these related works is given in [3].

3   WSOL Reusability Constructs

A large number of reusability constructs was developed for programming languages, from jumps and loops in assembly languages to hyperspaces and other concepts in modern languages. Of course, not all of them are applicable for a description language like WSOL. Before we have started development of WSOL, we have examined various existing reusability concepts and their usability in the context of WSOL.

We had two partially conflicting sets of goals. On the one hand, we wanted to achieve reusability of WSOL specifications and comprehensive descriptions of relationships between service offerings, as discussed in Section 1. On the other hand, our approach in designing WSOL was to provide solutions that are relatively simple to use and implement and lightweight in terms of run-time overhead. (For this reason, we have used in WSOL classes of service instead of more demanding custom-made SLAs, one language for various types of constraint and management statement instead of several specialized languages, support for management third parties, constraints evaluated periodically, and random evaluation of constraints [3].) To achieve simplicity and lightweightness of WSOL, we had to select a limited number of reusability constructs. Since WSOL was primarily developed for management of Web Services and their compositions, we were particularly interested in reusability constructs that support management applications, including dynamic adaptation [2].

We have studied a broad set of reusability constructs in programming and specification languages. For example, we have examined types, classes, single and multiple inheritance, polymorphism, overriding, restriction, sets and set operations (e.g., union, intersection, difference), scoping, domains, structs, templates, template instantiation, templatization, macros, include statements, subroutines, contracts, subcontracts, relationship tables, constraint dimensions, policies, roles, refinement, mixins, aspects, composition filters, hyperspaces, monitors, and views. In addition, we have examined three languages—Quality Interface Definition Language (QIDL) [12], Quality Modeling Language (QML) [13], and Quality Description Language (QDL) [14]—developed for the formal specification of QoS for Common Object Request Broker Architecture (CORBA) distributed objects, as well as reusability constructs in these languages.

We had no statistical data about the usage and usefulness of different reusability constructs in languages similar to WSOL. Therefore, we have tried to find some realistic practical examples that could justify inclusion of such constructs into WSOL. We have also tried to estimate how frequently such examples would occur. Further, we have tried to estimate the overhead of reusability constructs on WSOL and its tools. Another area that we have looked at was modeling of some constructs with other constructs, to reduce the total number of WSOL constructs. After determining options and alternatives, we have analyzed their costs and benefits and determined priorities for reusability constructs in WSOL.

As a result of our study, WSOL now contains the following **reusability constructs**:

1. The definition of service offerings (SOs).
2. The definition of constraint groups (CGs).
3. The definition and instantiation of constraint group templates (CGTs).
4. The extension (single inheritance) of service offerings, constraint groups, and constraint group templates.
5. The inclusion of already defined constraints, statements, and constraint groups.
6. The specification of applicability domains.
7. The declaration of operation calls.

We summarize explanations of these WSOL constructs in the following seven subsections. Additional explanations, discussions, and complete WSOL examples for all discussed constructs can be found in [5]. Some WSOL examples for reusability constructs are also given in the Appendix of this paper. In the Appendix examples, other elements of WSOL items are omitted for brevity. Precise syntax definitions (for an earlier version of WSOL) and further examples are given in [3].

In the eighth subsection, we briefly discuss the definition of external ontologies of QoS metrics, measurement units, and currency units used in WSOL files. The definition of such ontologies is not a part of the WSOL language, but these ontologies can be reused across WSOL service offerings of different Web Services.

3.1   Definition of Service Offerings (SOs)

A WSOL **definition of a service offering (SO)** contains formal definition of new constraints and statements that categorize the class of service represented with this service offering. It can also contain reusability constructs discussed in the following subsections. For example, constraints and statements that were already defined elsewhere can be included in a service offering using the WSOL inclusion reusability construct, discussed in Subsection 3.5. Further, a new service offering can be defined as an extension of an existing service offering, as discussed in Subsection 3.4. If inside one service offering two or more constraints of the same type (e.g., two pre-conditions) are defined for the same operation, they all have to be satisfied. This means that the Boolean 'AND' operation is performed between such constraints.

Every service offering has a name and exactly one accounting party. An accounting party is the management party responsible for logging all SOAP messages related to this service offering and for calculating monetary amounts to be paid [1]. We have recently added into WSOL optional specification of validity duration and/or expiration time of a service offering. If no validity duration or expiration time is specified, consumers can use the service offering until its deactivation or explicit switching of service offerings [2]. The Appendix contains example parts of the definition of the service offering 'SO9' for the 'buyStockService' Web Service. More detailed examples of the definition of WSOL service offerings can be found in [1–5].

In most cases, a WSOL service offering is specified for a particular Web Service. Different consumers of this Web Service can use the same service offering, potentially in parallel. However, if a service offering contains constraints and statements that do not reference particular ports and operations, it can be provided by different Web Services. This feature enables development of libraries of reusable service offerings. When the concept of a 'service type' becomes standard in WSDL, it will be straightforward to update WSOL with the definition of a reusable service offering specified for a WSDL service type.

3.2 Definition of Constraint Groups (CGs)

WSOL has a special reusability construct for the **definition of constraint groups**. A **constraint group (CG)** is a named set of constraints and statements. It can also contain reusability constructs, including definitions of other constraint groups. The number of levels in such **nesting of constraint groups** is not limited. (Since a constraint group can contain not only constraints, but also statements and reusability constructs, the name 'item group' would be more appropriate. We have kept the name 'constraint group' for compatibility with early versions of WSOL.) Analogously to service offerings, a new constraint group can be defined as an extension of an existing constraint group. In the earlier versions of WSOL, the Boolean 'AND' operation was always applied between constraints in the same constraint group. However, we have recently added constraint groups in which only some constraints have to be satisfied (at least one constraint or exactly one constraint). An example definition of the WSOL constraint group 'CG17' is shown in the Appendix.

While definitions of constraint groups and service offerings have some syntax similarities, they also have semantic and syntax differences. The crucial semantic difference is that consumers can choose and use service offerings, not constraint groups. Service offerings are used for the definition of complete and coherent offerings to consumers. On the other hand, the main goal of constraint groups is the easier development of WSOL files. They are used for smaller sets of constraints and statements. These sets can, but need not, be complete from the consumer viewpoint. An important syntax consequence is that accounting party, validity duration, and expiration time can be specified only for a service offering, not a constraint group. Further, dynamic relationships can be specified only for service offerings, not for constraint group. Another important syntax difference is that constraint groups can be nested, while service offerings must not be. In addition, while the Boolean 'AND' operation is always applied between constraints in a service offering, some other combinations of constraints can also be specified for constraint groups.

The WSOL concept of a constraint group has several benefits. First, a constraint group can be reused across service offerings as a unit, using the WSOL inclusion reusability construct discussed in Subsection 3.5. Second, it is possible to specify in a single management responsibility statement that all constraints from a constraint group are evaluated by the same management entity. Third, constraints in different constraint groups can have the same relative constraint name, so using constraint groups enables name reuse. Fourth, constraint groups can be used for different logical groupings of constraints and statements. For example, one can use a constraint group to group constraints and statements related to a particular port, port type, or operation. In this way, the concept of a constraint group complements the concept of a service offering that assembles constraints and statements on the level of a Web Service. In addition, one can use constraint groups to define aspects of service offerings. For example, one can group all functional constraints for one port type into one constraint group, QoS constraints for the same port type into another constraint group, and access rights for this port type into a third constraint group. This supports separation of concerns.

3.3 Definition and Instantiation of Constraint Group Templates (CGTs)

A **constraint group template (CGT)** is a parameterized constraint group. (Analogously to constraint groups, the name 'item group template' would be more appropriate.) This WSOL reusability construct is very useful because different classes of service often contain constraints with the same structure, but with different numerical values. WSOL contains two constructs for constraint group templates – one for their definition and the other for their instantiation. At the beginning of a **definition of a constraint group template**, one defines one or more abstract constraint group template parameters, each of which has a name and a data type. The definition of constraint group template parameters is followed by the definition of constraints, statements, and reusability constructs in the same way as for constraint groups. Constraints inside a constraint group template can contain expressions with constraint group template parameters.

An **instantiation of a constraint group template** provides concrete values for all constraint group template parameters. The result of every such instantiation is a new constraint group. One constraint group template can be instantiated many times, in different service offerings and for different Web Services. Two instantiations of the same constraint group template can provide different parameter values. The Appendix contains examples of the declaration of the constraint group template '*CGT2*' and its instantiation inside the service offering '*SO9*'. This constraint group template has one parameter: '*minAvail*' of the data type '*numberWithUnit*'.

The WSOL concept of a constraint group template currently has some limitations, adopted for simpler implementation. Constraint group templates cannot be defined inside service offerings, constraint groups, or other constraint group templates. They can be defined only as separate, non-contained, items in WSOL files. Also, since constraints inside a constraint group template may contain expressions with constraint group template parameters, these constraints must not be included inside other constraint group templates, constraint groups, or service offerings. In addition, WSOL does not support partial instantiation of a constraint group template. In such an instantiation, values for only some constraint group template parameters would be supplied and the result would be a new constraint group template.

3.4 Extension (Single Inheritance)

WSOL supports **extension (single inheritance) of service offerings, constraint groups, and constraint group templates** with their *'extends'* **attribute**. This enables easy definition of new WSOL items (service offerings, constraint groups, or constraint group templates) from existing ones. The extending WSOL item inherits all constraints, statements, and reusability constructs from the extended WSOL item and can define or include additional ones. We have also studied the use of multiple inheritance in WSOL, but decided not to support it due to complexities, such as the possibility of conflicts in accounting parties. Effects similar to multiple inheritance of constraint groups can be achieved by including multiple constraint groups inside a new constraint group [5]. An example of WSOL extension is given in the Appendix. The constraint group 'CG17' extends the constraint group 'CG6', which was defined in some other WSOL file.

Note that the extending WSOL items are not subcontracts (as defined in [15]) of the extended WSOL items. A subcontract weakens pre-conditions and strengthens post-conditions. This means that between the inherited and the additional pre-conditions the logical 'OR' operation is preformed, while for pre-conditions the logical 'AND' operation is preformed. In WSOL, the logical 'AND' operation is performed between the inherited and the additional constraints (unless these constraints are in a constraint group in which only some constraints have to be satisfied). This approach was chosen due to simplicity. It addresses reuse of WSOL items. For comparisons of WSOL service offerings, constraint groups, and constraint group templates, it would be useful to also support specification of subcontracts. We plan to address this issue in a future version of WSOL.

3.5 Inclusion

During definition of a new service offering, constraint group, or constraint group template, some of the constraints, statements, or constraint groups might have been already defined elsewhere. The WSOL **inclusion reusability construct** enables incorporating such constraints, statements, or constraint groups with a single XML element, instead of writing again their complete definitions. Reusable constraints, statements, and constraint groups can be defined once and included many times, across different service offerings, constraint groups, and/or constraint group templates, and even across different Web Services. Such reusable constraints, statements, and constraint groups can be defined inside service offerings, but also outside any service offering, e.g., in libraries of reusable WSOL items. In the Appendix, the WSOL inclusion construct is illustrated with the incorporation of the QoS constraint 'QoSCons2' (defined, in some other WSOL file, as part of some service offering 'SO1') into the constraint group 'CG17'.

We have recently built into WSOL optional naming of expressions, which are parts of constraints, and the possibility of inclusion of expressions into other expressions. This feature further improves reusability of WSOL constraints.

3.6 Specification of Applicability Domains

All WSOL constraints, statements, service offerings, constraint groups, constraint group templates, and operation calls are defined for some applicability domain. An **applicability domain** defines to which operations, port types, ports, and/or Web Services the given WSOL item applies. Declarations of used QoS metrics, discussed in Subsection 3.8, also contain specification of applicability domains, denoting for which operations, port types, ports, and/or Web Services the QoS metric is measured or computed.

In WSOL, an applicability domain is specified using the **attributes 'service'**, **'portOrPortType', and 'operation'**. Definitions of constraint group templates, constraint groups, statements, and non-periodic constraints, as well as declarations of operation calls and used QoS metrics have these three applicability domain attributes. However, definitions of service offerings and periodic QoS constraints have only the attribute '*service*'. This is because they do not describe an operation, port type, or port, but a complete Web Service.

We categorize applicability domains into specific and abstract. A **specific applicability domain** refers to a particular operation (or a group of operations) of a particular port (or a group of ports) of a particular Web Service. On the other hand, in an **abstract applicability domain**, the value of one or more of the applicability domain attributes refers to any operation, port type, port, and/or Web Service. An example of an abstract applicability domain is 'a particular operation of a particular port type (but not a particular port of a particular Web Service)'.

To represent special values of the applicability domain attributes, WSOL has built-in constants '*WSOL-ANY*', '*WSOL-EVERY*', '*WSOL-MANY*', and '*WSOL-ALL*' [4, 5]. '*WSOL-ANY*' is used in abstract applicability domains, while the latter three constants are treated as specific applicability domains. For example, the constraint group '*CG17*' from the Appendix has specific applicability domain, while its QoS constraint '*QoScons_3*' has abstract applicability domain.

The benefit of having abstract applicability domains is that WSOL items that have abstract applicability domains can be included (in the case of constraint group templates: instantiated) in different service offerings, constraint groups, and constraint group templates. This means that using abstract applicability domains is beneficial for reusability of WSOL items. On the other hand, to perform monitoring and management activities with WSOL service offerings, it is necessary to know precisely for which Web Service, port, and operation a WSOL constraint should be evaluated (or a QoS metric should be measured or computed). Consequently, using specific applicability domains is important for manageability of WSOL items. To conclude, the goals of reusable and manageable WSOL items conflict with each other.

To provide some support for reusability, while ensuring manageability of WSOL items, we have developed the concept of an actual applicability domain. The **actual applicability domain** of a contained WSOL item is calculated as an intersection of the values of its applicability domain attributes and the actual applicability domain of its immediate containing WSOL item (service offering, constraint group, or constraint group template). If the contained WSOL item is defined with an abstract applicability domain and the containing WSOL item has a specific actual applicability domain, then the actual applicability domain of the contained WSOL item can be specific. For example, the actual applicability domain of the '*QoScons_3*' within the '*CG17*' is specific. In this way, WSOL items can be both reusable and precise enough for management activities.

To ensure that applicability domains of containing WSOL items and their contained WSOL items produce meaningful actual applicability domains and do not conflict, we have defined rules for relationships between these applicability domains. In addition, we have defined rules for the specialization of domains during inclusion of items with abstract applicability domains. While these rules support both manageability and reusability of WSOL items, they also introduce additional complexity into WSOL. Detailed discussion and examples of these rules are given in [4, 5].

3.7 Declaration of Operation Calls

Expressions in WSOL constraints can contain calls to operations implemented by another Web Service, by the management entity evaluating the given constraint, or by the Web Service for which the constraint is evaluated. A **declaration of an operation call** enables referencing results of the same operation call (i.e., the same invocation) in different sub-expressions of one constraint or in several related constraints, without re-invoking the operation. Examples of the declaration of an operation call and its use in expressions can be found in [4, 5].

3.8 External Ontologies and WSOL Declarations of Used QoS Metrics

By '**definition of a QoS metric**' we mean 'all the information necessary to measure this QoS metric or compute it from other QoS metrics, as well as the information necessary to use this QoS metric'. For example, a QoS metric 'response time' can be computed as 'time immediately after the operation execution – time immediately before the operation execution' and its measurement unit can be 'second'.

An important reusability feature of WSOL is that QoS metrics used in WSOL specifications are not actually defined in WSOL files. They are defined in reusable **external ontologies** of QoS metrics [16]. WSOL also supports external ontologies of measurement units and currency units. Ideally, appropriate standardization bodies would define such ontologies and make them well-known. In practice, any other interested party (ranging from powerful multinational companies to interested individuals) can also define and publish such ontologies.

Once a QoS metric is defined in a reusable external ontology, its use for particular operations, port types, ports, and/or Web Services is declared in WSOL files. The Appendix shows the WSOL construct for the **declaration of a used QoS metric** '*DailyAvailability*' (defined in the ontology '*QoSMetricOntology*') inside the periodic QoS constraint '*QoScons4*' in the constraint group template '*CGT2*'.

Apart from the increased reusability, the outsourcing of definitions of QoS metrics, measurement units, and currency units into external ontologies can have other benefits [5]. First, the use of external ontologies eases comparisons between WSOL service offerings specified for the same or for different Web Services. If QoS metrics, monetary units, and currency units were defined within a service offering, comparisons of service offerings using the same QoS metrics, monetary units, and/or currency units would be harder. Second, the use of common external ontologies can decrease chance of semantic misunderstanding between provider Web Services and their consumers. Some QoS metrics, such as response time, can be defined in several ways [16]. Even when definitions of QoS metrics are precise, a semantic misunderstanding can occur if the involved parties have different interpretations of the basic terms used in these definitions. For unambiguous understanding of providers and consumers both precise definitions of QoS metrics and the use of common basic terminology are necessary. Ontologies can provide both. Third, such outsourcing can ease compilation of WSOL files and reduce run-time overhead of their monitoring, metering, and evaluation. This is because optimized code for monitoring and metering of most common QoS metrics (or at least some basic QoS metrics) from such ontologies can be built into the management infrastructure for WSOL.

We have summarized the requirements for ontologies of QoS metrics, measurement units, and currency units in [16]. The ontologies we currently use for our experiments with WSOL have very simple structure and only a few entries. We plan more work in this area, including the use of WSOL expressions in such ontologies, in the future.

4 Potential Influences of WSOL Reusability Constructs on Related Works

As overviewed in Section 2, there are several recent works on the formal and precise description of Web Services, their constraints (particularly QoS), and SLAs. Ideally, future standards for the comprehensive description of Web Services, in addition to WSDL, should integrate good features from all relevant works, including WSOL. We believe that WSOL reusability constructs (along with some other WSOL concepts and features, as discussed in [3]) can be a very useful input into such future standards. Some WSOL reusability constructs would also be a useful addition to the competing works. In this section, we outline potential influences of WSOL reusability constructs on WSLA, WSML, and WS-Policy, while further discussion—including potential influences on SLAng, DAML-S, and WSMF—is given in [5].

WSLA and WSML have relatively few reusability constructs. They both have some support for templates, but not as flexible as WSOL constraint group templates. In addition, WSLA has the concept of an obligation group (similar to the WSOL concept of a constraint group) and the concept of an operation group (more powerful than the specification of specific applicability domains in WSOL). Addition of other WSOL reusability constructs—particularly inclusion and extension (inheritance)—to these languages would enable easier development of WSLA and WSML specifications. Note also that WSLA and WSML files contain precise definitions of the QoS metrics, while we suggest outsourcing these definitions into reusable external ontologies. WSLA, WSML, and WSOL can be good starting points for the standardization of the language for such ontologies. On the other hand, the QoS metrics built into the SLAng schema can be used as input for the development of the ontologies.

The influences between reusability constructs in WSOL and WS-Policy can go in both directions. WS-Policy has several good reusability features, many similar to those present in WSOL. It enables definition of policies either inside definitions of subjects to which they refer (e.g., inside WSDL files) or independently from definitions of subjects. In the latter case, the policy and the subject are associated through an external attachment. An external attachment contains definition of an applicability domain, which contains an unordered set of subjects (e.g., WSDL services, ports, and/or operations). This definition of applicability domains is more flexible than the one used in WSOL, but it leaves greater space for semantic errors (e.g., when a policy containing references to message parts is attached to a subject where these message parts do not exist). Next, WS-Policy enables inclusion of parts of policies into other policies. This is analogous to the WSOL inclusion of constraint groups, constraints, statements, and expressions. Further, WS-Policy defines policy operators that group parts of policies. There are three policy operators in WS-Policy: *'All'*, *'ExactlyOne'*, and *'OneOrMore'*. Since they can be named and included as units, policy operators are similar to WSOL definitions of constraint groups. Our recent WSOL additions of constraint groups and constraint group templates in which only some constraints have to be satisfied and of naming and inclusion of expressions were influenced by WS-Policy.

On the other hand, some WSOL reusability constraints could be introduced into WS-Policy. In particular, policies and their parts may be parameterized, but WS-Policy does not provide further detail about this topic. We suggest extending WS-Policy with a precise definition of templates, similar to WSOL constraint group templates. Also, supporting inheritance, in addition to inclusion, would enable easier comparison of policies. Some other extensions of WS-Policy, such as precise specification of management information for the specified policies, are also needed [3].

5   Conclusions and Future Work

WSOL has a diverse set of reusability constructs: definition of service offerings (SOs), definition of constraint groups (CGs), definition and instantiation of constraint group templates (CGTs), extension, inclusion, specification of applicability domains, and declaration of operation calls. These constructs enable defining common or similar parts of WSOL specifications once and using these definitions many times. Consequently, they provide easier specification of new WSOL items from existing WSOL items for the same Web Service or different Web Services.

We see practical applications for all WSOL reusability constructs. For example, the definition of abstract applicability domains enables defining reusable WSOL items (service offerings, constraint groups, constraint group templates, constraints and/or statements) that can be shared between various Web Services through inclusion and constraint group template instantiation. In particular, constraint group templates can be instantiated many times, with different parameter values. Since significant similarities can exist even for classes of service of different Web Services, constraint group templates are particularly reusable WSOL items.

Ideally, an appropriate standardization body (or at least, some powerful companies) would define most common and most reusable Web Service description items (e.g., WSDL port types, WSOL items) and make them well-known. Such libraries of reusable Web Service description items would complement reusable ontologies of QoS metrics, measurement units, and currency units. If such reusable libraries and ontologies were defined and used widely in practice, the heterogeneity of Web

Services would be alleviated since it would be easier to compare Web Services and their classes of service, as discussed next.

WSOL reusability constructs determine static relationships between service offerings, which show similarities and differences between service offerings. Consequently, they can be used in the process of selection of Web Services and their service offerings. Such static relationships complement dynamic relationships between service offerings, specified outside definitions of service offerings. The process of comparison, negotiation, and selection of Web Services, their functionality, and particularly QoS is a very complex issue, without a simple and straightforward solution. WSOL static and dynamic relationships between service offerings can guide this process, but appropriate algorithms and heuristics have to be developed. Further, static relationships between service offerings also support the mechanisms for dynamic (i.e., runtime) manipulation of classes of service discussed in [2], particularly dynamic creation of new classes of service.

Several recent related works—WSLA, WSML, WS-Policy, SLAng, DAML-S, and WSMF—address issues that partially overlap with WSOL. In some aspects, they are more powerful than WSOL. However, WSOL also has its advantages [3], such as the richer set of reusability constructs. Future standards for the comprehensive description of Web Services (additional to WSDL) should integrate good features from all these works, including WSOL. Some WSOL reusability constructs would also be a useful addition to the competing works.

While the focus of our current and future research is on the WSOI management architecture, we are also working on enhancing the WSOL language. In particular, we have recently extended WSOL in several ways, some of which are related reusability constructs. One example is the addition of constraint groups and constraint group templates in which only some constraints have to be satisfied. Another example is the possibility of naming and inclusion of expressions. These and some other recent WSOL improvements (e.g., the more consistent and more versatile syntax for WSOL statements) will be explained and illustrated in a forthcoming publication. We also plan some other minor additions to reusability constructs in future versions of WSOL. An example is subcontracting of service offerings, constraint groups, and constraint group templates, as discussed in Section 3.4. Other improvements of WSOL, such as specification of service fees for third-party management and accounting parties, are also envisioned. Nonetheless, WSOL is relatively complete and stable and its reusability constructs have both the expressive power and significant practical applications. They can positively influence the competing works and future standards in this area.

References

1. Tosic, V., Pagurek, B., Patel, B. Esfandiari, B., Ma, W.: Management Applications of the Web Service Offerings Language (WSOL). In Proc. of the 15th Conference On Advanced Information Systems Engineering – CAiSE'03 (Velden, Austria, June 2003). Lecture Notes in Computer Science (LNCS), No. 2681. Springer-Verlag (2003) 468–484
2. Tosic, V., Ma, W., Pagurek, B., Esfandiari, B.: On the Dynamic Manipulation of Classes of Service for XML Web Services. In Proc. of the 10th Hewlett-Packard Open View University Association (HP-OVUA) Workshop (Geneva, Switzerland, July 2003). Hewlett-Packard (2003)

3. Tosic, V., Patel, K., Pagurek, B.: WSOL – A Language for the Formal Specification of Classes of Service for Web Services. In Proc. of ICWS'03 – The First International Conference on Web Services (Las Vegas, USA, June 2003). CSREA Press (2003) 375–381

4. Patel, K.: XML Grammar and Parser for the Web Service Offerings Language. M.A.Sc thesis, Carleton University, Ottawa, Canada. Jan. 30, 2003. On-line at: http://www.sce.carleton.ca/netmanage/papers/KrutiPatelThesisFinal.pdf (2003)

5. Tosic, V., Patel. K., Pagurek, B.: Reusability Constructs in the Web Service Offerings Language (WSOL) [Second Extended Revision]. Res. Rep. SCE-03-21, Department of Systems and Computer Engineering, Carleton University, Ottawa, Canada. (Sep. 2003)

6. Ludwig, H., Keller, A., Dan, A., King, R.P., Franck, R.: Web Service Level Agreement (WSLA) Language Specification, Version 1.0, Revision wsla-2003/01/28. International Business Machines Corporation (IBM). On-line at: http://www.research.ibm.com/wsla/WSLASpecV1-20030128.pdf (2003)

7. Sahai, A., Durante, A., Machiraju, V.: Towards Automated SLA Management for Web Services. Res. Rep. HPL-2001-310 (R.1), Hewlett-Packard (HP) Labs Palo Alto. July 26, 2002. On-line at: http://www.hpl.hp.com/techreports/2001/HPL-2001-310R1.pdf (2002)

8. Lamanna, D.D., Skene, J., Emmerich, W.: SLAng: A Language for Defining Service Level Agreements. In Proc. of the 9th IEEE Workshop on Future Trends in Distributed Computing Systems - FTDCS 2003 (Puerto Rico, May 2003). IEEE-CS Press (2003) 100–106

9. Hondo, M., Kaler, C. (eds.): Web Services Policy Framework (WS-Policy), Version 1.0. Dec. 18, 2002. BEA/IBM/Microsoft/SAP. On-line at: ftp://www6.software.ibm.com/software/developer/library/ws-policy.pdf (2002)

10. The DAML Services Coalition: DAML-S: Semantic Markup for Web Services. WWW page for DAML-S version 0.7. Oct. 2, 2002. On-line at: http://www.daml.org/services/daml-s/0.7/daml-s.html (2002)

11. Fensel, D., Bussler, C.: The Web Service Modeling Framework WSMF. Internet white paper and internal report, Vrije Unversiteit Amsterdam. On-line at: http://informatik.uibk.ac.at/users/c70385/wese/wsmf.paper.pdf (2002)

12. Becker, C., Geihs, K., Gramberg, J.: Representing Quality of Service Preferences by Hierarchies of Contracts. In Proc. of the Workshop Elektronische Dienstleistungswirtschaft und Financial Engineering – FAN'99 (Augsburg, Germany, 1999). Schüling Verlag (1999)

13. Frolund, S., Koistinen, J.: Quality of Service Specification in Distributed Object Systems Design. In Proc. of the 4th USENIX Conference on Object-Oriented Technologies and Systems - COOTS '98 (Santa Fe, USA, Apr. 1998), USENIX (1998)

14. Zinky, J.A., Bakken, D.E., Schantz, R.E.: Architectural Support for Quality of Service for CORBA Objects. John Wiley & Sons, Theory and Practice of Object Systems, Vol. 3, No. 1 (Apr. 1997)

15. Meyer, B.: Applying "Design by Contract". IEEE, Computer, Vol. 25, No. 10 (Oct. 1992) 40-51

16. Tosic, V., Esfandiari, B., Pagurek, B., Patel, K.: On Requirements for Ontologies in Management of Web Services. In Proc. of the Workshop on Web Services, e-Business, and the Semantic Web at CAiSE'02 (Toronto, Canada, May 2002). Lecture Notes in Computer Science (LNCS), No. 2512. Springer-Verlag (2002) 237-247

Appendix: Examples of Some WSOL Reusability Constructs

```
<wsol:CG name = ìCG17î extends = ìSO1ns:CG6î service =
ìbuyStock:buyStockServiceî portOrPortType = ìWSOL-MANYî
operation = ìWSOL-MANYî >
   Ö
  <wsol:include constructName = ìSO1ns:SO1.QoScons2î />
  <wsol:constraint name = ìQoScons_3î xsi:type =
ìqosSchema:QoSconstraintî service = ìWSOL-ANYî portOrPortType
= ìWSOL-EVERYî operation = ìWSOL-EVERYî >
    Ö
   </wsol:constraint>
  Ö
</wsol:CG>
<!-- ~~~~~~~~~~~~~~~~~~~~~~~~~~~~~~~~~~~~~~~~~~~~~~~~~~~~~~ -->
<wsol:CGT name = ìCGT2î service = ìWSOL-ANYî portOrPortType =
ìWSOL-MANYî operation = ìWSOL-MANYî >
  <wsol:parameter name= ìminAvailî dataType =
ìwsol:numberWithUnitî unit = ìQoSMeasOntology: Percentageî />
  Ö
  <wsol:constraint name = ìQoScons4î xsi:type =
ìPeriodQosSchema:PeriodicQoSconstraintî service = ìWSOL-ANYî >
    Ö
    <expressionSchema:booleanExpression>
      <expressionSchema:arithmeticWithUnitExpression>
        <expressionSchema: QoSmetric metricType =
ìQoSMetricOntology: DailyAvailabilityî service = ìWSOL-ANYî
portOrPortType = ìWSOL-ALLî operation = ìWSOL-ALLî
measuredBy = ìWSOL_INTERNALî />
      </expressionSchema:arithmeticWithUnitExpression>
      <expressionSchema:arithmeticComparator type = ì>=î />
      <expressionSchema:arithmeticWithUnitExpression>
        <expressionSchema: arithmeticWithUnitVariable
aWUName = ìtns:CGT2.minAvailî />
      </expressionSchema:arithmeticWithUnitExpression>
    </expressionSchema:booleanExpression>
  </wsol:constraint>
  Ö
</wsol:CGT>
<!-- ~~~~~~~~~~~~~~~~~~~~~~~~~~~~~~~~~~~~~~~~~~~~~~~~~~~~~~ -->
<wsol:serviceOffering name = ìSO9î service = ìbuyStock:
buyStockServiceî accountingParty = ìWSOL-SUPPLIERWSî >
  Ö
  <wsol:instantiate CGTName = ìtns:CGT2î resService =
ìbuyStock:buyStockServiceî resPortOrPortType = ìWSOL-MANYî
resOperation = ìWSOL-MANYî resCGName = ìCG19î >
    <wsol:parmValue name = ìminAvailî >
      <wsol:numberWithUnitConstant>
        <wsol:value> 95 </wsol:value>
        <wsol:unit type = ìQoSMeasOntology: Percentageî />
      </wsol:numberWithUnitConstant>
    </wsol:parmValue>
  </wsol:instantiate>
  Ö
</wsol:serviceOffering>
```

Event Based Web Service Description and Coordination

Wilfried Lemahieu, Monique Snoeck, Cindy Michiels, Frank Goethals,
Guido Dedene, and Jacques Vandenbulcke

Katholieke Universiteit Leuven, Department of Applied Economic Sciences,
Naamsestraat 69, B-3000 Leuven, Belgium
{wilfried.lemahieu, monique.snoeck, cindy.michiels,
frank.goethals, guido.dedene,
jacques.vandenbulcke}@econ.kuleuven.ac.be

Abstract. This paper proposes the concept of *business events* as the corner-stone to web service description and coordination. First, a web service architecture is introduced as the result of an event based analysis & design phase. Then, it is advocated how the event concept can be used for semantically rich web service description. A distinction is made between two web service interfaces: a non-transactional *query interface* and a transactional *event notification interface*. Furthermore, a web service *composition model* is proposed, based on *event broadcasting* and *event preconditions*, instead of traditional one-to-one method invocations. The composition model is presented in a static variant and in a version with dynamic subscription. Throughout the paper, it is shown how the event based approach fits entirely within the current standard SOAP/WSDL/UDDI web services stack.

1 Introduction

The web services concept can be considered as a revolutionary paradigm for loosely coupled application integration within and across enterprise boundaries. It promises to bring about a revolution in the way business partners can integrate their information systems, allowing for innovative organizational forms that were unthinkable before [1]. A web service can be looked upon as a *public, remote interface* to certain functionality, where the actual implementation is *hidden* from the applications that use it. In this aspect it is very similar to distributed object technologies such as RMI, CORBA and DCOM, which are, however, restricted to the intranet. In contrast, web services use a lightweight XML messaging protocol, SOAP [2], which is applicable across the entire Internet, without being hampered by companies' firewalls.

Still, despite its obvious great promises, the web service paradigm hasn't fully lived up to its expectations (yet), at least not at the *inter-enterprise* level. Indeed, web services are well established for intra-enterprise application integration (EAI) and even *static* business-to-business interaction (B2Bi), i.e. in an extended enterprise with fixed, long-standing business partners. However, they fail to provide more than very basic services at the level of *dynamic* B2Bi, where services dynamically find one another and enter ad-hoc partnerships to perform complex business transactions. Current implementations of dynamic B2Bi are largely limited to rather simple request/response services such as currency converters, stock information services, i.e. non transactional systems. In this respect, web service technology didn't succeed in

C. Bussler et al. (Eds.): WES 2003, LNCS 3095, pp. 120–133, 2004.
© Springer-Verlag Berlin Heidelberg 2004

fulfilling two of its primary promises: fully automated *discovery* and *invocation* of (remote) services and fully automated *composition* of "atomic" services into ad-hoc complex, transactional systems.

This paper advocates how failure to achieve those two goals can, at least partially, be imputed to the one-to-one interaction paradigm that inherently underlies "traditional" SOAP messaging. As an alternative, an approach to web service interaction is proposed, based on the simultaneous participation in (and processing of) *shared business events*. Event notifications are not propagated one-to-one but are *broadcast* in parallel to all services that have an interest in an event of the corresponding type. Yet, this broadcasting paradigm is fully compatible with current web service standards such as SOAP, WSDL and UDDI. The paper is structured as follows: Section 2 briefly overviews how the business event concept can be used at the level of *analysis*, *design* and *implementation* of web services. In Section 3, business events are used to enhance *web service description*, distinguishing between a (non-transaction) query interface and a (transactional) event notification interface. Section 4 discusses *web service composition and coordination*, again through participation in shared events. Conclusions are formulated in Section 5.

2 Event Based Web Service Development

The event based interaction mechanism as proposed in this paper directly reflects an underlying object-oriented analysis and design methodology: MERODE [3,4]. MERODE is complementary to UML [5], which can then be used as a *formalism* to capture the MERODE specifications. MERODE represents an information system through the definition of business events, their effect on enterprise objects and the related business rules. Although it follows an object-oriented approach, it does not rely on "pure" method invocation to model interaction between domain object classes as in classical approaches to object-oriented analysis, e.g. [6]. Instead, *business events* are identified as independent concepts. An object-event table (OET) allows defining which types of objects are affected by which types of events. When an object type is involved in an event, a method is required to implement the effect of the event on instances of this type. Whenever an event actually occurs, it is broadcast to all involved domain object classes. For example, let us assume that the domain model for an order handling system contains the four object types CUSTOMER, ORDER, ORDERLINE and PRODUCT. The corresponding UML Class diagram is given in Fig. 1.

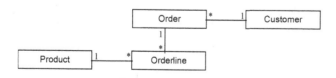

Fig. 1. Domain model for an order handling system

Business event types are e.g. *create_customer*, *modify_customer*, *create_order*, *ship_order*, etc. The object-event table (see Table 1) shows which object types are affected by which types of events and also indicates the type of involvement: C for creation, M for modification and E for terminating an object's life. For example, *create_orderline* creates a new occurrence of the class ORDERLINE, modifies an occurrence of the class PRODUCT because it requires adjustment of the stock-level of the ordered product, modifies the state of the ORDER to which it belongs because the number of outstanding order lines has to be increased, and modifies the state of the CUSTOMER of the order because the total cost of outstanding orders has to be updated.

Table 1. Object-event table for the order handling system

	CUSTOMER	ORDER	ORDERLINE	PRODUCT
create_customer	C			
modify_customer	M			
end_customer	E			
create_order	M	C		
modify_order	M	M		
end_order	M	E		
customer_sign	M	M		
ship_order	M	M		
Bill	M	M		
create_orderline	M	M	C	M
modify_orderline	M	M	M	M
end_orderline	M	M	E	M
create_product				C
modify_product				M
end_product				E

The event based domain model is combined with a *behavioral model*. Each enterprise object type has a method for each event type in which it may participate. Such method specifies *preconditions* put on the corresponding event type by the object type and implements an object's *state changes* (i.e. changes to attribute values) as the consequence of an event of the corresponding type. For example, when a customer orders a product, a new ORDERLINE is created, which involves the following business events: *create_order* and *create_orderline*. Preconditions may be based on *class invariants* (such as attribute constraints and uniqueness constraints) and on *event sequence constraints* that can be derived from a finite state machine associated with the object type. For example, Fig. 2 shows a finite state machine for the ORDER domain object. As long as it is not signed, an order stays in the state "existing". The *customer_sign* event moves the order into the state "registered". From then on the order has the status of a contract with the customer and it cannot be modified anymore: the events *modify_order*, create_orderline, modify_orderline and end_orderline are no longer accepted for this order. The *ship_order* event signals that the order has been shipped to the customer, after which the order object reaches the "shipped" state. Finally, the *bill* event signals the billing of the order. In this way, the sequence constraints mimic the general business process(es). Full details of how to construct such an object-event table and validate it against the data model and the behavioral model are beyond the scope of this paper but can be found in [3,4].

The entirety of all enterprise objects that together shape the business process(es) is called the *enterprise layer*. The eventual information system is realized as a layer on top of the enterprise layer, consisting of output and input services. *Output services* use attribute inspections to query the enterprise objects and deliver the information to

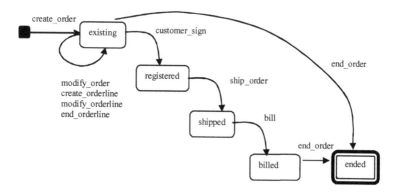

Fig. 2. State machine for an ORDER object

the user. Upon occurrence of a business event in the real world, *input services* collect input data from the user and invoke the corresponding event to update the set of enterprise objects. This enterprise layer (and the associated input and output services) can be implemented in multiple ways: standalone or distributed, tightly coupled or loosely coupled. Currently, existing implementations are built around e.g. *stored procedures* [4] or an *EJB framework* [7].

As discussed in detail in [8], an implementation by means of web service technology can be achieved in three stages, in line with the Model Driven Architecture of the OMG [9]. First, the business rules are captured into a MERODE-based, technology neutral *business model*, which defines the enterprise objects and their interaction through participation in shared business events. The business model is then "enriched" into an *architectural model*, which groups the enterprise objects into distributed, loosely coupled components: the actual web services. Here, attribute inspection and event broadcasting are identified as distinct, complementary interaction types. In a third stage, the architectural model is translated into an actual technology-bound *implementation model*, based on current state-of-the-art technologies such as SOAP and WSDL.

A generic architectural model for a web services environment is depicted in Fig. 3. The enterprise layer is distributed across different services, with each service controlling/consisting of a subset of the enterprise objects. A web service is conceived as a layered structure: its local *enterprise layer* incorporates the actual business logic. This layer also contains *stub objects*, which locally represent external web services. The layer above is called the *event layer* and consists of an *event dispatcher*, which also has a *transaction management* task. The highest layer is the *interface layer*. A web service can be accessed through two types of interfaces: a *query interface* and an *event notification interface*. Separate *input services* and *output services*, combined with a *user interface*, allow for human users to interact with the web service. For a thorough discussion of the proposed architecture, we refer to [8]. Throughout the remainder of this paper, the relevant components of the architecture will be clarified. The next section deals with both types of interfaces, as they will be the foundation to event based web service description.

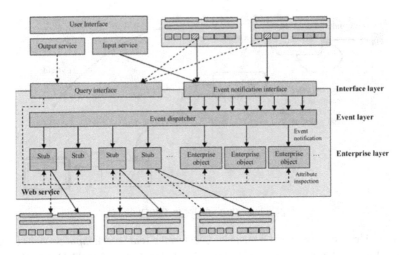

Fig. 3. A generic architectural model for a web services environment

3   Event Based Web Service Description

3.1 Introduction

A key requirement for a successful web service is that potential users are able to find the right service and obtain the information necessary to interact with it. The way in which web services can be advertised and discovered strongly resembles the CORBA approach: CORBA objects are described by their *IDL interface*, which can be published in an *IDL repository* [10]. WSDL [11] and UDDI [12] can, to a certain extent, be considered as web service variants of respectively IDL and IDL repositories. However, whereas an intranet-based technology such as CORBA still pertains to a manageable number of services, transposing a similar approach to a world-spanning environment such as the Web may result in very poor searching performance. Specifying a service solely in terms of its *interface*, i.e. its input and output message types as is done in WSDL only offers a very incomplete picture, especially since the ultimate goal of many web services is to provoke changes in the *real world*, e.g. debiting a credit card in exchange for the delivery of a book at a certain address. This *business logic*, i.e. what the service "does" may be a much more valuable search criterion. Although UDDI provides a categorization mechanism according to "real-world" criteria such as industry branch, product type and geographic location, this only accommodates for a rough, first-level filtering of available services. It is in no way destined at discovering a service based on fine-grained specifications of what is required from it. In this section, we discuss how the event based approach can make a web service's interface itself more descriptive by discerning between two interfaces: a *query interface* and an *event notification interface*. Further on, we indicate how the interface based description can be complemented with information about a service's *process logic*.

3.2 The Query Interface

The query interface allows reading the attribute values from the enterprise objects that make out a web service. Attribute inspection is a rather straightforward type of interaction and is based on a one-to-one process. The interaction does not cause a state change: the enterprise objects' attributes are never "written to" through a web service's query interface. Therefore, preconditions resulting from class invariants or sequence constraints are not applicable to attribute inspections[1], nor should such invocation be subject to a transaction context. In its simplest form, the query interface publishes a set of public *getAttribute()* methods at the service level, which are mapped transparently to *getAttribute()* methods on individual enterprise objects. Obviously, performance can be boosted if multiple correlated attribute inspections are bundled into a single SOAP message exchange. Therefore, in a more complex form, the query interface may allow invoking real *query methods* that inspect multiple enterprise objects in the web service and which may also calculate aggregate values, check for the existence of a certain object etc. In the latter case, a query method's input parameters behave as selection criteria that denote to which object(s) the attribute inspections apply.

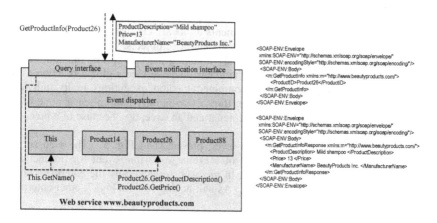

Fig. 4. Invocation on a web service's query interface

Fig. 4 presents the simplified example (at instance level) of the query interface for a "BeautyProducts Inc." manufacturer web service. Invoking the *GetProductInfo()* method on the query interface results in the appropriate attributes being inspected on several enterprise objects, as determined by the method's input parameter "Product26". The requested values are communicated in a single return message. The example includes a (simplified) SOAP message that represents an invocation to this interface, along with a return message that contains the query result.

---

[1] The sole exception may be checks for the appropriate access privileges, which is beyond the scope of this paper.

3.3 The Event Notification Interface

The situation where a service A "writes" to a service B, i.e. causes attribute values in B's enterprise objects to be updated, is a bit more complex because the updates that result from a given business event are to be coordinated throughout the entirety of all enterprise objects that participate in the event. These combined updates must be considered as a single transaction. A service is never allowed to directly invoke *setAttribute()* methods on another service's enterprise objects. A service can only "write to" another service by inducing a *business event* on this service, which may affect the state of one or more enterprise objects embedded in the service, provided that all constraints are satisfied.

An event is induced by invoking the appropriate method on a web service's event notification interface. If a relevant event occurs in the real world, e.g. a customer issues a purchase order, a stock drops below a threshold value, an order is shipped, ... this event is acknowledged by an input service, e.g. by a user pressing the "sign" button in a sales order form. The input service notifies the web service by an invocation on the web service's event notification interface. For each event type "understood" by the service, the event notification interface has a separate method. Attributes that describe the event (e.g. order quantity) are passed as input parameters to the method.

Upon invocation of such method, the event notification interface passes control to the *event dispatcher*. The latter implements a "local" OET and knows which event types are relevant to which (types of) local enterprise objects. The event is *broadcast* by the event dispatcher by simultaneously invoking the appropriate method on each enterprise object that participates in the event. The service's global reaction to the business event corresponds to the combined method executions in its individual enterprise objects. The corresponding transaction is only committed if none of the objects that take part in the event have generated an exception because of a precondition violation.

An example is, again at instance level, presented in Fig. 5. The web service for "BeautyProducts Inc." is notified of a *create_orderline* event for product #26, order #56 and with a quantity of 30. In *this* web service, only the enterprise object "Product26" is affected by the event. As will be depicted in Fig. 7, other enterprise

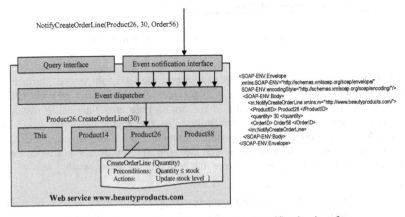

Fig. 5. Invocation on a web service's event notification interface

objects in other web services can also be involved. The event is broadcast by the event dispatcher to Product26. The latter updates its stock level, provided that the constraint "order_quantity ≤ stock" is satisfied. Included in the example is a sample SOAP message that represents a corresponding event notification. Note that, in contrast to traditional method based interaction, no return values are sent other than possibly a status variable for transaction management purposes, i.e. to inform the event dispatcher whether the event was accepted or whether it failed because a constraint was violated.

3.4 Discovery and Invocation of Web Services

A "traditional" SOAP based method invocation on a web service may have two effects: it may cause a state change in the web service (as part of the *execution* of the method) and it may retrieve information from the service (by means of its *return value*). The *event* paradigm differs from this standard method invocation approach in that the cases of "*reading from*" a web service and "*writing to*" a web service are strictly separated. The latter is called the *Command-Query Separation* design principle, which, as discussed in general in [13], makes software much more transparent and makes classes easier to (re)use. [14] discusses its design and maintenance advantages specific to a web services context.

However, the distinction between a query interface and an event notification interface also allows for richer web service description. The query interface describes which information can be provided by the service. The interface not only *describes* the service, it can also be *invoked* by search engines to use the web service's current state as part of the selection criteria when searching for a web service, e.g. to search for an on-line library, which currently has the title ("XML for dummies") available.

In contrast, the event notification interface denotes to which types of real world events the service will respond, hence provides information about *what the service does*. The event paradigm has the advantage of reconciling a business concept with a programming concept: business events such as purchases, out-of-stock events etc. but also events in the programming sense of the word, i.e. to which services can *subscribe*. This business aspect of events does not reduce the description of an event notification interface to mere a software concept, but also links it to "real world events".

Both a web service's query interface and event notification interface are fully compatible with current web service standards. "Traditional" web services or search engines will be confronted with a standard WSDL interface description. An example of a (partial) WSDL description of the BeautyProducts service's interface is depicted in Fig. 6.

Once the appropriate web service has been selected, one needs to be able to retrieve information about how a request to the service is to be conducted. Traditionally, this will include details of the input parameters that are to be provided and the output parameters that can be expected. The latter will suffice for very simple services, e.g. a currency conversion service, but it may prove to be limiting in the case where multiple types of inputs and intermediate conditions may result in a diverse range of possible output types, e.g. an online trip booking service. In such case, an explicit description of the service's *logic* may be required. Also, "real world" properties, which cannot be considered input or output parameters but which may definitely affect the outcome of a transaction, should be taken into consideration. As to these,

```
<definitions name="BeautyProductsService"
xmlns:s="http://www.w3.org/2001/XMLSChema/"
xmlns:s0="http://www.beautyproducts.com/"
targetNamespace="http://www.beautyproducts.com/">
  <types>...</types>
  <message name="GetProductInfoSoapIn">
      <part name="parameters" element ="s0:GetProductInfo"/>
  </message>
  <message name="GetProductInfoSoapOut">
      <part name="parameters" element ="s0:GetProductInfoResponse"/>
  </message>
  <message name="NotifyCreateOrderLineSoapIn">
      <part name="parameters" element ="s0:NotifyCreateOrderLine"/>
  </message>
  <portType name="ManufacturerPortType">
      <operation name="GetProductInfo">
        <input message="s0:GetProductInfoSoapIn"/>
        <output message="s0:GetProductInfoSoapOut" />
      </operation>
      <operation name="NotifyCreateOrderLine">
        <input message="s0:NotifyCreateOrderlineSoapIn"/>
      </operation>
  </portType>
  <binding>...</binding>
  <service>...</service>
</definitions>
```

Fig. 6. Example of a partial WSDL description for the BeautyProducts web service

the event based approach can again be utterly useful, simply because the event notion bridges the gap to real world concepts.

Taking this one step further, one could imagine the web service's *state machine* being published as metadata to the service, as applied in DAML-S [15]. An event may induce a state transition to one or more of the service's objects (and, as discussed further on in this paper, to one or more services). The published state machine would be a unified version of the respective state machines of the internal enterprise objects, retaining only these states and transitions that are relevant to the outside world. In this way, the calling application would know in advance which preconditions are required for a certain event to be accepted and which results (i.e. postconditions) can be assured. In contrast, an invocation to the query interface will never induce a state transition.

Finally, such event based description lends itself well to be enhanced with semantic web concepts. Ontology languages such as OWL [16] can be used to describe the "meaning" of a certain business event type, parameters, states etc. at a semantic level, as advocated in [17].

4   Event Based Web Service Coordination, Composition and Choreography

4.1   A Composition Model for Static B2Bi

As already stated, the web service concept hasn't fulfilled its full potential (yet) at the level of complex, composite services. Standards for web service *transactions*, *composition* and *choreography* are still under development. Existing attempts such as BTP [18], WSCL [19] BPEL4WS [20] or BPML [21] are based on quite divergent assumptions. This paper does not try to introduce yet another standardization proposal. Rather, it discusses how the event broadcasting concept may provide a composition model that facilitates the development of a standard.

In our opinion, an important factor in the troublesome process of web service coordination and choreography is the fact that SOAP, as it is traditionally used, is essentially a *one-to-one mechanism*. The latter may be adequate for simple request/response services, but is inherently difficult to co-ordinate in a complex environment where numerous business partners interact in shared business processes. With an event based paradigm, transactional interaction, i.e. a "write" operation is essentially a *broadcasting* mechanism instead of a one-to-one method invocation: event broadcasting is *implemented* by means of SOAP method invocations, but the methods *are executed in parallel and in a coordinated way on all enterprise objects that participate in the event*. In this way, a single business event results in multiple simultaneous updates in multiple enterprise objects. Previous sections discussed this issue at the level of a single service. This section denotes how events can mould coordinated interactions *between* services, which are much easier to model than the myriad of one-to-one message exchanges that could make out a single business transaction in a pure method based approach.

Each web service may have its own local input services, which generate local business events. These events can be dispatched to the service's local enterprise objects that participate in the event, based on a "local" object event table. As to remote objects to which the event may be relevant, the proposed architecture caters for an explicit *subscription* mechanism. In this way, a given web service's event dispatcher will dispatch its events to all local objects *and* to all remote services that are subscribed to the corresponding event type. For that purpose, a web service's architectural model is enriched with *stub objects* that locally "represent" an individual remote service and that contain a reference to it, e.g. its URL. The general purpose of these stub objects is to make the distribution aspect of the enterprise layer transparent to a web service's local enterprise objects and to its event dispatcher. A first function of a stub object is to "mirror" attributes that belong to the enterprise objects of the remote service it represents (see [8] for more details). However, more importantly, it also *propagates* event notifications to the remote service it represents. As illustrated in Fig. 7, the resulting event based interaction mechanism takes place in four stages: if an event occurs in a given web service, as initiated by an invocation to its event notification interface (1), this event is broadcast to all appropriate *local* objects, i.e. enterprise objects and stub objects (2). Each stub object propagates the event by invoking the appropriate method on the event notification interface of the external service it represents (3). Such remote service's event dispatcher then in its turn broadcasts the event to its own local objects (4). Some of these may also be stub objects that further propagate the event etc. The appropriate enterprise objects each execute a corresponding method, in which preconditions are checked and/or updates are executed (5).

Stub objects play an important role in distributed transaction management and web service choreography. Indeed, within an individual service, the event dispatcher enforces the "atomicity" of an event: an event only succeeds if no preconditions are violated in any of the enterprise objects that participate in it. The event dispatcher commits the transaction if no "reject" messages were returned by any local enterprise object. Local stub objects may also reject an event: a stub object does not impose preconditions itself, but receives an accept/reject return value upon invocation to

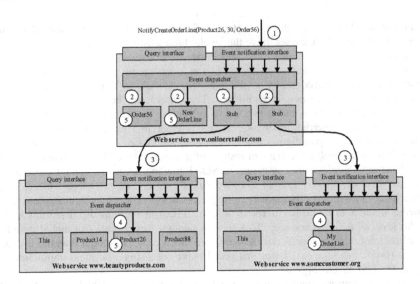

Fig. 7. Example of event propagation by means of stub objects

the event notification interface of the remote web service it represents. If the (partial) transaction fails in the remote service (because preconditions were not met in one or more of the latter's own enterprise objects), the stub receives a "reject" return value. As a consequence, the stub rejects the event locally and in its turn returns a reject value to the local event dispatcher. In this way, the concept of event broadcasting and propagation results in a distributed transaction mechanism. Whereas the underlying SOAP invocations only represent one-to-one interactions, events can be seen as coordinated actions that affect multiple services. They are initially induced on a single service, but are propagated to all services subscribed to the event (event subscription is discussed in section 4.2) and finally result in parallel method invocations on all (local and remote) enterprise objects that participate in the event. The action is only committed if no constraint are violated in any of the participating services/enterprise objects. In this way, events shape a coordination mechanism, derived from the business model, over the individual SOAP invocations.

In this way, event propagation can be used both for interaction between peer services (by the stubs propagating event notifications) and for a complex service to coordinate the behavior of its components (by the local event dispatcher notifying local objects). This approach can be easily generalized into an n-level system: each local object and each remote service may in their turn be complex objects, which react to the event by further propagating it to their constituent objects or (through their own stubs) to yet other remote services etc. The event is propagated recursively at each level in a hierarchy of complex web services with a complex task, that in their turn are conceived of more simple services with a simpler task until a level of atomic services is reached. The choreography is not stored in a central document: the appropriate consecution(s) of events is/are determined by the preconditions imposed by the individual enterprise objects. These preconditions emanate from the state machines of the enterprise objects, hence reflecting the actual business processes.

4.2   A Composition Model for Dynamic B2Bi

The previous section actually dealt with a model for *static* B2Bi. In such situation, all interacting partners "know" one another in advance: together they forman extended enterprise. In such situation, dynamic event subscription is not necessary; stub objects are created at the moment when the interacting web services are deployed.

This approach can be extended into a composition model for *dynamic* B2Bi, where partners have to dynamically *find* one another, after which they participate in short lived, ad hoc partnerships. In that case, stub objects will be created at runtime, when a certain remote service is selected for interaction. Just like the "real" enterprise object types, stub object types are represented in the object-event table. The latter denotes which event types have a "create" effect on a stub object type. If such an event is induced, a new stub object instance is created (at least if all preconditions are satisfied) and initialized with the URL of the external service it represents. This URL is to be provided as one of the event's parameters. From then on, the stub object is responsible for the interaction between the two services. In this way, dynamic subscription of one service to another, i.e. the initiative to start interacting, is modeled as part of the business processes and embedded in the lifecycle of the enterprise objects. In a similar way, events with an "end" effect on a stub object terminate the interaction between two services.

In this respect, one last issue is how the "decision" of one service to search for and subscribe to another service is to be made. This paper does not address the search (and possible matchmaking) mechanism itself, but its contribution lies in the ability to automatically formulate web service search criteria through the concept of *failed events*. Indeed, an event can be seen as a transaction. Web transactions typically have transaction management mechanisms that are less strict than the traditional "ACID" properties from the database world. Especially, "long-lived" transactions (e.g. the online booking of a trip) will not always be committed or rolled back in their entirety, but will consist of sub-transactions for which an alternative (i.e. a "corrective" action) has to be sought, if one of them fails. In the context of events, a failed transaction can be translated as a precondition not being satisfied in one of the enterprise objects (either local or remote). Instead of rolling back the entire event, a measure of *goal seeking* intelligence can be added to the event dispatcher so as to come up with a corrective action for the failed precondition, instead of aborting the entire transaction. Two alternatives can be discerned, which can be enhanced with semantic web based mediation capabilities as discussed in [22]. A first possibility is that the event dispatcher of the web service where the event was initially induced tries to induce (an) additional, "subordinate" event(s) on the service where the event was refused. The subordinate event(s) should result in state changes that resolve the original precondition violation. In this respect, the precondition can be seen as the *goal* that is to be achieved to be able to have the original event succeed. The subordinate events can be selected based on whether their *postconditions* assist in achieving this goal. For example, a *purchase* event in an on-line shop could be rejected because it is "membership required". The calling service could then try to induce a *new_membership* event, so as to satisfy the precondition for the original purchase event. A second option is to simply "replace" the service where the precondition was violated with a similar service that imposes less strict preconditions. In that case, the new service will be selected based on the events it understands *and* on the preconditions it imposes on them. For example, the calling service could look for another shop, which does not require membership. Again, search criteria not only entail interface formats, but also specifications about the effect(s) on the real world.

5 Conclusions

A rigorous analysis and design phase is often overlooked with respect to web service development [23]. This paper proposed a business modeling paradigm where an information system's behavior is closely related to real world *business events*. The relation between static and dynamic specification is made by means of a very elegant and concise modeling technique: the object event table. The event based specification can be *implemented* without giving up on the current de facto standard web services stack based on SOAP, WSDL and UDDI.

Moreover, the clear distinction between query interface and event notification interface results in improved web service description capabilities. The description can be further enhanced by an explicitation of the *business logic* by means of state machines, with business events inducing the state transitions.

Events also facilitate the coordination of interacting web services. Discerning attribute inspections from event notifications allows for focusing on only the latter with respect to transaction management. Also, the fact that event propagation is a broadcasting mechanism yields a composition model that is much simpler than streamlining the myriad of one-to-one message exchanges in a purely RPC based approach.

Not unlike [24], statecharts are used to capture the business processes, rather than flowcharts as applied e.g. in [25]. Particular to our approach is that *business events* are a core modeling component, triggering and coordinating state transitions in multiple enterprise objects and services. The "choreography" is distributed as preconditions imposed by the respective enterprise objects and services that participate in an event; at each level the behavior of a composite system is the union of the individual objects' behavior. Dynamic B2Bi is facilitated by incorporating a goal seeking and subscription mechanism, again based on business events and preconditions resulting from the state machines that reflect the underlying business processes.

References

1. Leclerc, A.: Distributed Enterprise Architecture, Integrating the Enterprise, Vol. III, no. 5 (2000)
2. Seely, S., Sharkey, K.: SOAP: Cross Platform Web Services Development Using XML, Prentice Hall PTR, Upper Saddle River, NJ (2001)
3. Snoeck, M., Dedene, G.: Existence Dependency: The key to semantic integrity between structural and behavioral aspects of object types, IEEE Transactions on Software Engineering, Vol 24 No. 24, pp. 233–251 (1998)
4. Snoeck, M., Dedene, G., Verhelst M, Depuydt A. M.: Object-oriented Enterprise Modeling with MERODE, Leuven University Press, Leuven (1999)
5. OMG, Unified Modelling Language, http://www.omg.org/uml/
6. Jacobson, I., Christerson, M., Jonsson, P. et al.: Object-Oriented Software Engineering, A use Case Driven Approach, Addison-Wesley, Reading MA, Rev. 4th pr. (1997)

7. Lemahieu, W., Snoeck, M., Michiels C.: An Enterprise Layer Based Approach to Application Service Integration, Business Process Management Journal, (forthcoming)
8. Lemahieu, W., Snoeck, M., Michiels, C., Goethals, F., Dedene, G., Vandenbulcke, J.: A Model Driven, Layered Architecture for Web Service Development, K.U.Leuven – F.E.T.E.W. research report (2003)
9. OMG, Model Driven Architecture, http://www.omg.org/mda/
10. Siegel, J.: CORBA – fundamentals and programming, John Wiley & Sons, New York, NY (1996)
11. Web Services Description Language (WSDL) 1.0. specification, http://msdn.microsoft.com/xml/general/wsdl.asp (2001)
12. UDDI Technical White Paper, Ariba, IBM Corporation and Microsoft Corporation, (2000)
13. Meyer, B.: Object-oriented software construction, 2nd edition, Prentice Hall PTR, (1997)
14. Lemahieu, W., Snoeck, M., Michiels, C., Goethals, F.: An Event Based Approach to Web Service Design and Interaction, accepted for APWeb'03, LNCS 2642 (2003)
15. DAML-S official website, http://www.daml.org/services/daml-s/2001/05/ (2001)
16. Dean, M., Schreiber G., Van Harmelen, F. et al.: OWL Web Ontology Language 1.0 Reference, W3C Working Draft (2003)
17. Lemahieu, W.: Web service description, advertising and discovery: WSDL and beyond, in: Vandenbulcke J. and Snoeck M. (eds.): New Directions In Software Engineering, Leuven University Press, Leuven (2001)
18. Potts, M., Cox, B., Pope, B.: Business Transaction Protocol Primer, OASIS Committee Supporting Document (2002)
19. Banerji, A. et al.: Web Services Conversation Language (WSCL) 1.0, W3C Note (2002)
20. Weerawana, S., Curbera, F.: Business Process with BPEL4WS, IBM white paper (2002)
21. Arkin, A.: Business process Modeling Language, BPMI draft specification (2002)
22. Fensel, D., Bussler, C., Ding, Y., Omelayenko, B.: The Web Service Modeling Framework WSMF, Electronic Commerce Research and Applications 1(2), (2002)
23. Frankel, D., Parodi, J.: Using Model-Driven Architecture to Develop Web Services, IONA Technologies white paper (2002)
24. Benatallah, B., Dumas, M., Sheng, Q., Ngu, A.: Declarative composition and peer-to-peer provisioning of dynamic web services, ICDE 2002 (2002)
25. Casati, F., Jin, L., Ilnicki, S., Shan, M.C., An open, Flexible and Configurable System for Service Composition, HPL Technical report HPL-2000-41 (2000)

A New Web Application Development Methodology: Web Service Composition[*]

Zhihong Ren, Beihong Jin, and Jing Li

Institute of Software, Chinese Academy of Sciences, Beijing 100080, China
{ren, jbh, lij}@otcaix.iscas.ac.cn

Abstract. Traditional methodology for web application development can not fully satisfy the requirement raised by web services. Generally, distributed web applications are built on top of 3-tier client/server model, which is a relatively static environment that can provide reliable static service binding. However, in the web service-oriented environment, service binding is dynamic and just-in-time. How to enable the composition among the web services has become a key area in the software engineering research. In this paper, we present a new methodology for web application development, which is a framework that facilitates the visual design, validation and development of web service composition. The framework is mainly based on Web Service Composition Graph (WSCG), the underlying formalism for web service compositions. Using graph grammar and graph transformation defined on WSCG, the static topological structure of a web service composition can be described and the automation of the constructed web service composition is also facilitated.

1 Introduction

In the past few years, web application design and implementation models are based on technologies that do not provide abstractions for capturing high-level design concepts. Therefore it is difficult to track design models in the implementation [8]. How to maintain the consistency between design models and implementation becomes a difficult task. On the other hand, with the development of the Internet and XML technologies, the more web services from different service providers becoming available on the Internet; the more web application scene will be dominated by web services in non-mission critical environments over the next several years. The traditional web application development methodology does not fulfill the requirements of web service-oriented environment, such as service dynamic composition and just-in-time binding [12]. Web services are becoming a common way of publishing e-business applications, because web service has certain advantages: firstly, web services participated in are loosely coupled. Secondly, the interactions among the web services use the messages exchange in standard XML format (mostly the candidate is

---

[*]This work is partially supported by the National Natural Science Foundation of China under Grant No. 60173023; the National High Technology Development 863 Program of China under Grant No. 2001AA414020.

C. Bussler et al. (Eds.): WES 2003, LNCS 3095, pp. 134–145, 2004.

the messages exchange in standard XML format (mostly the candidate is SOAP). Lastly, the most advantage of web service is that it is easy to compose web services to complex web applications.

Web-based business environments have become exceedingly dynamic and competitive in recent years. However, web application development technology does not yet provide enough flexibility to support the dynamic nature of web applications. In order to build web applications that are based on web services, it is important to model the dependencies among web services. The first step is to represent composition relationships among web services. The next step is to provide operational semantics to the composite model, which guarantees the sequences of interactions among web service composition. Current research devoted very little attention to these issues [10].

Based on business processes, web service composition refers to combining two or more web services to achieve the desired business goal. In this paper, we introduce the web services composition framework, which facilitates the visual design, development and execution of web service composition consistently and also supports the service dynamic composition and just-in-time binding. Our work is mainly based on Web Service Composition Graph (WSCG), the underlying formalism for web services composition. Using graph grammar and transformation [13] defined on WSCG, the static topological structure of a web service composition can be described and the execution of the constructed composite web services is supported. Graph transformation simulates the evolution of the overall structure of large-scale web applications. At the same time, the binding between composition models and web service operations can be established either at design time (static composition) or at run-time through WSCG rules (dynamic composition).

The rest of the paper is organized as follows. In section 2, some brief related works and the background of web service composition are discussed. Section 3 outlines the components of web service composition framework. Section 4 gives the basic concepts of WSCG. Some preliminaries about graph grammar and graph transformation are also presented in this section. Meta-model and operational semantics of web service composition are emphasized. Section 5 details the architecture and key technology of web service composition supporting system. Section 6 concludes this paper and describes our further work.

2 Background and Related Works

In contrast to traditional web applications, web services are application building blocks that use XML for information interchange. Simple Object Access Protocol (SOAP), Web Services Description Language (WSDL) and Universal Description, Discovery and Integration (UDDI) are common standards [5], which enable creation, deployment, description, discovery and communication among web services. A framework for composing web services must be adherent to these standards. There have been several alternative languages that are specific the interaction between web services and guarantee of the SOAP messages delivery. These proposals that enable the orchestration of web services include Web Services Flow Language (WSFL) [9],

XLANG [14] and Business Process Execution Language for Web Services (BPEL4WS) [4].

However, these specification languages are of textual forms without adequate semantics constraints. Specifications written in text-based languages are difficult to understand, verify and visualize. Furthermore, there are important features missing in these languages: graphical representation that is flexible to design web service compositions and dynamic web service composition, which is important to reflect the loose coupling and scalable mediation of web services in a service-oriented architecture. Lack of enough semantic constraint, it is difficult to design these specifications, even the structural errors in specifications cannot be detected. Despite some specifications, such as WSFL, provides some graphical notions to illustrate web service composition, but lacks precise definitions and semantics of the graphical notions and thus does not insure the consistency between graphical representation and specification of executable web services composition.

Several research efforts model web service compositions using Petri net [7, 11] and activity diagram [1, 15]. But grammar-directed web service composition and dynamic adaptation of transformation productions (rules) are not yet supported. Furthermore, graph grammar can be regarded as a proper generalization of Petri nets [2], where the state of a system is described by a graph instead of a collection of tokens. Other than activity diagram, arbitrary constraints can be added into composition model through defining additional constraint productions.

3   Web Service Composition Framework

Our focus is to establish a reliable link among abstract specification, refined design, verification and the ultimate execution of web applications based on web services. We propose a framework consisting of the following components (as illustrated in Figure 1).

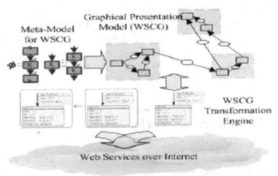

Fig. 1. Web service composition framework

1)   Graphical Presentation Model. Describing web service composition with text alone is ineffective and prone to errors. We introduce WSCG as a graphical presentation for web service composition; a WSCG is a directed, acyclic and attributed graph whose nodes represent web services participating in web service composition and whose edges describe the control and data link among nodes.

2) Meta-Model for Composition. Distributed web application development is becoming an activity of composing web services into ultimate applications. Rules are needed to deal with sequential, concurrent and etc. behaviors. We use WSCG grammar to describe how web services are assembled into compositions that represent business logics at the intuitive (graphical) level. A meta-model for composition is a set of productions for web services composition during design. Directed by WSCG grammar, user can design a well-formed web service composition, whose links between the imported web services are well regarded and whose overall structure satisfies reachability, liveness and deadlock-freedom.

3) WSCG Transformation Engine. The meta-model for composition addressed above is not limited to the graphical web service composition. Assigning operational semantics of WSCG, web service composition denoted by WSCGs are mode executable. WSCG transformation is a set of productions that identify the WSCG operational semantics. WSCG transformation engine invokes and orchestrates web service composition through interpreting a WSCG and dynamically maintaining WSCG instances. WSCG transformation engine also makes web service composition dynamic and automated. Dynamic web service composition is the ability to discover and use remote services just in time. Web service composition automated can be accomplished through engine without requiring human intervention.

4 Web Service Composition Model

We firstly introduce the basic concepts of graph grammar and transformation [6, 13] and then give the formal WSCG definition,

4.1 Preliminaries

Definition 4.1 **Labeled Graphs**, Given two fixed alphabets Ω_V and Ω_E for node and edge labels, respectively, a labeled graph over (Ω_V, Ω_E) is a tuple $G = (G_V, G_E, s_G, t_G, lv_G, le_G)$, where G_V is a set of nodes, G_E is a set of edges, $s_G, t_G: G_E \rightarrow G_V$ are the source and target functions, and $lv_G: G_V \rightarrow \Omega_V$ and $le_G: G_E \rightarrow \Omega_E$ are the node and the edge labeling functions, respectively. A *total graph morphism* $f: G \rightarrow G'$ is a pair $f = (f_V: G_V \rightarrow G'_V, f_E: G_E \rightarrow G'_E)$ of functions which preserve sources, targets and labels, i.e. which satisfies $f_V o\ t^G = t^{G'} o f_E, f_V o\ s^G = s^{G'} o f_E, lv^{G'} o f_V = lv^G$ and $le^{G'} o f_E = le^G$. A *subgraph* S *of* G, *written* $S \subseteq G$, *is a* labeled graph *with* $S_V \subseteq G_V, S_E \subseteq G_E$, $s^S = s^G|_{s_E}, t^S = t^G|_{s_E}, lv^S = lv^G|_{s_V}$ and $le^S = le^G|_{s_E}$. A *partial graph morphism* g from G to H is a total graph morphism from some subgraph $dom(g)$ of G to H, and $dom(g)$ is called the domain of g.

As for introducing formally the concept of an attributed graph, we need use some basic notions of universal algebra [6]. Attributes are labels of graphical objects taken from attribute algebra. Hence, an attributed graph consists of a labeled graph and an attribute algebra, together with some attribute functions connecting the graphical and the algebraic part.

Definition 4.2 Attributed Graph, Given a label alphabet Ω and a signature $Sig = (S, OP)$ [6]. then $G = (G_V, G_E, s^G, t^G, lv^G, le^G, G_A, av^G, ae^G)$ is a Sig-attributed graph, where: $G = (G_V, G_E, s^G, t^G, lv^G, le^G)$ is an Ω-labeled graph. G_A is a Sig-algebra. av^G: $G_V \rightarrow U(G_A)$ and ae^G: $G_E \rightarrow U(G_A)$ are the node and the edge attributing functions, respectively.

Definition 4.3 Production, Graph Grammar, A *production* P: L $\xrightarrow{r}$ R consists of a production name P and of an injective partial morphism r in **Alg(Sig)** [6]. The graphs L and R are called the left- and the right-hand side of P, respectively. We also make reference to a production P: L $\xrightarrow{r}$ R simply as L $\xrightarrow{r}$ R. A *graph grammar* $\hat{G}$ is a tuple $\hat{G} = ((r_p)_{p \in P}, G_0)$ where $(r_p)_{p \in P}$ is a family of production morphisms indexed by production names, and G_0 is the start graph of the graph grammar. For attributed graph, if L and R have a common *subgraph* K, the following restrictions are fulfilled: i) The sources and targets of common edges are common nodes of L and R, i.e. $\forall\, e \in K_E \Rightarrow s^L(e) = s^R(e) \wedge t^L(e) = t^R(e)$. ii) Common edges and nodes of L and R do not differ with respect to their labels in L and R, i.e. $\forall\, e \in L_E \cap R_E \Rightarrow le^L(e) = le^R(e) \wedge \forall\, v \in L_V \cap R_V \Rightarrow lv^L(v) = lv^R(v)$.

Using productions of WSCG grammar; we can easily and intuitively describe web service composition. At above, we can only specify when these transformations should occur according to positive application conditions, which concerns the existence of certain nodes, edges and attributes of them. It is possible to specify negative application conditions for each particular production. The general idea of negative application conditions is to have a left-hand side not only consisting of one graph but also of several ones, connected by morphisms $L \xrightarrow{l} \hat{L}$, called constraints, with original left-hand side L. For each constraint, $\hat{L} - l(L)$ represents the forbidden structure. A match satisfies a constraint if it cannot be extended to the forbidden graph $\hat{L}$.

Definition 4.4 Application conditions
1)   A *negative application condition*, or *application condition* for short, over a graph L is a finite set A of total morphisms $L \xrightarrow{l} \hat{L}$, called constraints.
2)   A total graph morphism $L \xrightarrow{m} G$ *satisfies* a constraint $L \xrightarrow{l} \hat{L}$, written $m \models l$, if there is no total morphism $\hat{L} \xrightarrow{n} G$ such that $n \circ l = m$. m satisfies an application condition A over L, written $m \models A$, if it satisfies all constraints $l \in A$.
3)   An application condition A is said to be consistent if there is a graph G and a total morphism $L \xrightarrow{m} G$ satisfies A.

A production with application condition $\hat{p}$: $(L \xrightarrow{m} R, A(p))$, or condition production for short, is composed of a production named $\hat{p}$, a pair consisting of a partial morphism p and an application condition $A(p)$ over L. It is applicable to a graph G at $L \xrightarrow{m} G$ if m satisfies $A(p)$. In this case, the direct derivation $G \overset{p,m}{\Rightarrow} H$ is called direct conditional derivation $G \overset{p,m}{\Rightarrow} H$.

4.2 Web Service Composition Graph

Definition 4.5 Web Services Composition Graph (WSCG), Let $\Omega = (\Omega_V, \Omega_E)$ be a pair of label alphabets for nodes and edges and a signature $Sig = (S, OP)$. A WSCG G $= (G_V, G_E, s^G, t^G, lv^G, le^G, G_A, av^G, ae^G)$ is a Sig-attributed graph, where:

1) $\Omega_V = \{atomic, nesting, iterative\}$ and $\Omega_E = \{data, atomic, and\text{-}split, and\text{-}join, xor\text{-}split, xor\text{-}join\}$. We denote $\Omega = \Omega_V \cup \Omega_E$.

2) $S = \{string, address, mesg, condition, validity, wscg\}$. The *strings* are used for nodes and edges Names. The sort of *address* is used for web service location (URI). The *mesg* is used for ImportMesg and ExportMesg. The *condition* sort is used for PreCondition, PostCondition and TransCondition, while the *validity* sort is used for indicating TargetValid of edge transition condition. Finally, *wscg* is a reference of subgraph of WSCG.

3) G_V is a set of nodes, G_E is a set of edges, $s_G, t_G: G_E \to G_V$ are the source and target functions, and $lv_G: G_V \to \Omega_V$ and $le_G: G_E \to \Omega_E$ are the node and the edge labeling functions, respectively.

4) G_A is a Sig-algebra. $av_G: G_V \to U(G_A)$ and $ae_G: G_E \to U(G_A)$ are the node and the edge attributing functions, respectively.

According to the definition of WSCG, nodes and edges of WSCG can be labeled both fixedly and mutably to define WSCG. Fixed labels are node and edge types used for WSCG structuring. Mutable labels are attributes used to store data related to graphs. A type can be a string and attributes, which are specified by an attribute tuples consisting an attribute name, an attribute (data) type and an attribute value.

4.3 WSCG Grammar - Meta Model for Web Service Composition

The meta-model of web service composition is a set of WSCG grammars, which form the guideline for constructing web service composition, represented as WSCGs. Figure 2 (I) shows a slice of the graph grammar for WSCG, containing productions used for WSCG structure construction. The Start production replaces a $\varnothing$ graph by two *terminal* nodes (start and end nodes, denoted by the lowercase letters s and e respectively) and one *nonterminal* node (denoted by the capital letter K, I *and* N), which are connected through directed edges. If one node (or edge) is typed as "*", it means that this node (or edge) can be matched by any type node (or edge). The Sequential production describes the basic flow structure and defines the sequential execution order of the web service nodes occurring in the left-hand side. The Parallel production is used to describe concurrent flow within a WSCG, where only nonterminal nodes connected by neither O-S nor O-J labeled edges can be replaced with a pair of nodes. The Choice production is used to build an alternative mutually exclusive flow in WSCG, where only nonterminal nodes connected by neither A-S nor A-J labeled edges can be replaced with a pair of nodes. The Iterative production replaces a K node with I node, which represents the repetition of a group of web services until the exit condition is fulfilled. The Nesting production supports the notion of sub-flow, by which the execution of N nodes will trigger the execution of a subgraph of the WSCG.

Figure 2 (II) shows the attributing WSCG grammar containing the productions that for attributes of nodes and edges in a WSCG. These productions ensure that, after being applied, a WSCG will be well formed by the WSCG definition.

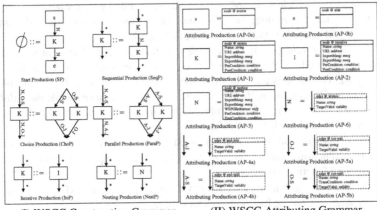

(I) WSCG Construction Grammar   (II) WSCG Attributing Grammar

Fig. 2. Meta-model for web service composition

There are five node attributing and five edge attributing productions. AP-0a and AP-0b attribute the source node and sink node, respectively. AP-1 replaces a nonterminal nodes K by an atomic node attributed with name, URI, ImportMesg, ExportMesg, PreCondition and PostCondition and the K is discarded. AP-2 similar to AP-1 except that Ap-2 replaces I node with an iterative node. AP-3 replaces an N node with a sub-WSCG.

Five edge attributing productions are defined, one for each edge type. AP-5b is defined for an edge with an O-S label, the O-S edge should be labeled an additional attribute of TransCondition. All other edges are attributed with name and TargetValid. We will give the informal operational semantics of WSCG in terms of attributes that specifies how to interpret a WSCG as follows.

4.4 WSCG Transformation – Operational Semantics for Web Service Composition

Figure 3 shows all transformation productions for WSCG interpretation. According to definition 4.3, match of the left hand side of production includes: structure of WSCG, node and edge types and node and edge attributes' values. In Figure 3, if one node (or edge) is typed as "*", it means that this node (or edge) can be matched by any type node (or edge).

Transformation production-1a initiates WSCG interpretation instances by modifying the TargetValid value of outgoing edges to "*true*". In the end, a WSCG interpretation instance is finished if only transformation production-1b is applied. TP-2a, 2b and 2c, these three productions, deal with all other incoming edges of the WSCG interpretation instance. Production TP-2c, which contains a negative application condition, can be applied to an and-join typed node only if this node does not contain any other incoming edge with *"false"* value in TargetValid attribute. Application of TP-2c acts as a synchronizer of all and-join typed edges. If any type of node (typed as "*") with an ingoing edge is typed with atomic, TP-2a will be applied to set the value of Pre-Condition to *"true"*. If with an ingoing edge is typed with xor-join, TP-2b will be matched. At this point, we need point out that WSCGs generated by WSCG grammar G do not contain a node with different types of incoming (or outgoing) edges, which is the reason why productions in Figure 3 are enough to interpret these WSCGs. All other outgoing edges are interpreted by TP-3a, 3b and 3c. If PostCondition attribute of a node is *"true"*, their atomic (and and-split) outgoing edges' TargetValid attributes will be set to "true" by TP-3a (and 3b). When the outgoing edge is typed with xor-split, there is some different of that what TargetValid attribute is set according to the expression of TransCondition attribute. Finally, three types of nodes are interpreted by transformation production-4a, 4b and 4c. PostCondition attribute of a node can be set to *"true"* only if the execution of represented web service is finished, i.e. result of these web services is obtained. PostCondition value of an atomic typed node's is directly set to "true" after the corresponding web service is performed, shown as TP-4a. If the node is typed with nesting, its PostCondition value can be set to *"true"* after total WSCG subgraph it references to is finished. TP-4c deals with iterative node. Until the expression of PostCondition is evaluated to *"true"*, the web service does not execution.

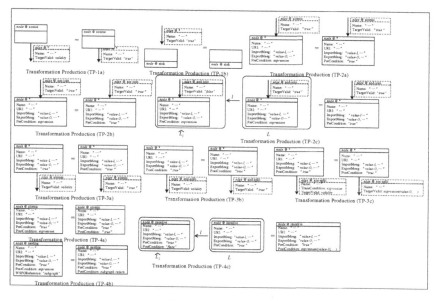

Fig. 3. WSCG operational semantics

When a new interpretation instance of WSCG is created, the TargetValid attributes of all edges connected with the start node are set to "true". Then, according to the TargetValid values of all incoming edges connected to this node, the PreCondition field of the node is evaluated. Only if its PreCondition value is "true", a node can be computed, i.e., the web service represented by the node can be invoked. If a node is typed with atomic, upon the completion of the corresponding web service execution, the PostCondition field is set to "true". When a node is typed with nesting, the Post-Condition expression is evaluated according to the parameters returned from the execution of the corresponding subgraph. For a node typed with iterative, the web service is not repeated until the PostCondition expression is evaluated to "true". If a node's PostCondition has been set to "true", the connected outgoing edges' TargetValid fields are determined. If the outgoing edge without a TransCondition field, the TargetValid field is simply set to "true"; otherwise, the value of the TargetValid field of the edge is determined by the TransCondition expression. Only if all the web services contained in a WSCG completed execution, the TargetValid fields of incoming edges of the end node are set to "true". At this point, execution of an instance of the WSCG is completed.

5   Web Service Composition Supporting System

Let us firstly consider a concrete example to show how to compose web services at design stage. And then the supporting system for web service composition at run-time is introduced.

5.1   Web Service Composition at Design Stage – Case Study

Figure 4 shows a web application based on WSCG. It is an e-Payment example. Customer can issue checks according to the requirement of e-Ticks web services. A customer issues a payment request to an agent when a check is needed. Then the agent accesses account in different banks according to the request. If foreign exchange is required, the agent may check bank-2; otherwise the agent only needs to check bank-1. After obtaining the digital signature from the manager, the agent issues the check required by the customer.

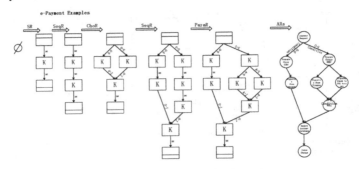

Fig. 4. Example of web service composition using WSCG grammar

Figure 4 also shows how to specify a web application into web service composition using WSCG grammar. First, the initial rule is applied by which we get a composite web service with start and end nodes. Subsequently, the sequential, choice, sequential and parallel rules are applied to the WSCG. The topology of web service composition is dynamically established.

5.2   Web Service Composition at Run-Time – Supporting System

Based on WebGOP [3], a prototype of web service composition supporting system is implemented, and several demo applications are developed on the prototype. In this section we describe the overall architecture of the supporting system and discuss several key issues in implementing the prototype.

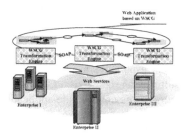

Fig. 5. WSCG based web application environment

Figure 5 shows the WSCG-based web applications. The basic building blocks of this infrastructure are the WSCG transformation engines, which provide containers for WSCGs and take charge of running a group of web services represented by the WSCGs. The engine is responsible for maintaining the context of each node and edge by associating with them. WSCG transformation engines interact with each other via SOAP messages, which ensure that the framework is compatible with the underlying web service environment. The topology of the whole WSCG is shared by all participating WSCG transformation engines. The state of the whole WSCG is co-managed by all subgraphs. The behavior of each WSCG node is defined by the web service bound to it. As discussed above, a web application consists of a set of subgraphs. During execution, the WSCG transformation engines coordinate with one another by invoking the operations specified by the WSCG. At the same time, WSCG transformation engines invoke the remote web services according to the specifications of WSCG subgraphs.

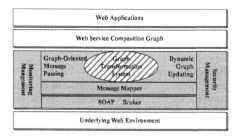

Fig. 6. WSCG transformation engine architecture

The architecture of WSCG transformation engine (grey area), as shown in Figure 6. The core of WSCG transformation engine is a set of graph-oriented message passing primitives, dynamic graph updating primitives, a graph transformation system, a message mapper and a SOAP broker. A set of graph-oriented message passing primitives and a set of basic dynamic graph updating primitives are provided for WSCG transformation. The graph-oriented message passing primitives include synchronous and asynchronous sends and receive functions for *unicast*, *multicast* and *anycast* communications. The dynamic graph updating primitives modify the topology of internal WSCG by adding/removing nodes/edges and changing the attributes of WSCG's nodes and edges, such as binding/unbinding web services to/from nodes and rewriting the nodes and edges' attribute values with the execution results from remote web services. Based on these primitives and WSCG graph, graph transformation system applies the WSCG interpretation productions in Figure 3. The Message mapper constructs SOAP messages for web service invocations and primitive transports, and parses the corresponding messages. The SOAP broker sends and receives invocation and result messages.

Reliable and flexible security is essential to the application of the WSCG transformation engine in open network environment. An authentication mechanism based on cryptic communication and digital signature identification should be implemented as an integral part of the engine. In order to help the developer to manage, debug and monitor their applications, a system monitoring management is required.

6 Conclusions and Future Works

In the web service-oriented environment, there are several attempts to create web applications using standards and protocols. In this paper, we present web service composition as a new methodology for web application development. Graph grammar-based web service composition helps to develop consistent web applications in dynamic service binding and loose-coupled web environment. At the same time, structural errors are eliminated from the target WSCG through using WSCG grammars. A framework, which facilitates the visual design, validation and development of web service composition, is also proposed. Using WSCG grammar and WSCG transformation productions, the static topological structure of a web service composition can be described and the automation of the constructed web service composition is achieved. Future work should define the semantics of web services graph transformation rules based on business rules.

References

1. Aiello, M., Papazoglou, M.-P., Yang, J., Carman, M., Pistore, M., Serafini, L., Traverso, P.: A Web Service Planning Language for Service Composition, In Buchmann, A., Casati, F., Fiege, L., Hsu, M.-C., Shan, M.-C. (Eds.): *Technologies for E-Services (TES 2002)*, LNCS, Vol. 2444. Springer-Verlag (2002), 76–85.
2. Bardohl, R., Ermel, C., Padberg, J.: Formal Relationship between Petri Nets and Graph Grammars as Basis for Animation Views in GenGED, *Proceeding of International Conference on Design and Process Technologies (IDPT 2002)*, Pasadena, USA (2002).

3. Cao, J., Ma, X., Chan, T.S., Lu, J.: WebGOP: A Framework for Architecting and Programming Dynamic Distributed Web Applications, *Proceedings of the 2002 International Conference on Parallel Processing*, Vancouver, B.C., Canada (2002).

4. Curbera, F., Goland, Y., Klein, J., Leymann, F., Roller, D., Thatte, S., Weerawarana, S.: Business Process Execution Language for Web Services (BPEL4WS), http://www-106.ibm.com/developerworks/library/ws-bpel/ (2002).

5. Curbera, F., Duftler, M., Khalaf, R., Nagy, W., Mukhi, N. Weerawarana, S.: Unraveling the Web Services Web: An Introduction to SOAP, WSDL, and UDDI, *IEEE Internet Computing*, Vol.6, No.2 (2002), 86–93.

6. Ehrig, H., Heckel, R., Korff, M., Löwe, M., Ribeiro, L., Wagner, A., Corradini, A.: Algebraic Approach to Graph Transformation, In: G. Rozenberg (Eds.): *Handbook of Graph Grammars and Computing by Graph Transformation*, World Scientific Publishing (1997), 247–312.

7. Hamadi, R., Benatallah, B.: A Petri Net-based Model for Web Service Composition, In: Schewe, K.-D., Zhou, X.-F. (Eds): *Conferences in Research and Practice in Information Technology*, Vol.17, Adelaide, Australia, Australian Computer Society, 191–200.

8. Koehler, J., Tirenni, G., Kumaran, S.: From Business Process Model to Consistent Implementation: A Case for Formal Verification Methods, *Proceeding of the Sixth International Enterprise Distributed Object Computing Conference*, Switzerland(2002), 96–108.

9. Leymann, F.: Web Service Flow Language (WSFL1.0), http: //www-4.ibm.com/ software/solutions/webservices/pdf/WSFL.pdf, May 2001.

10. Leymann, F., Roller, D., Schmidt, M.T.: Web Services and Business Process Management, *IBM System Journal*, Vol.41, No.2 (2002)

11. Mecella, M., Presicce, F.P., Pernici, B.: Modeling E-service Orchestration through Petri Nets, In Buchmann, A., Casati, F., Fiege, L., Hsu, M.-C., Shan, M.-C. (Eds.): *Technologies for E-Services (TES 2002)*. LNCS, Vol. 2444. Springer-Verlag, (2002) 38–47.

12. Stal, M.: Web services: beyond component-based computing, *Communications of the ACM*, Vol.45, No.10 (2002) 71–76.

13. Taentzer, G.: A Visual Modeling Framework for Distributed Object Computing, In: V.B. Jacobs and A. Rensink (eds.): *Formal Methods for Open Object-based Distributed Systems*, Kluwer Academic Publishers (2002).

14. Thatte, S., "XLANG: Web Services for Business Process Desgin", http:// www.gotdotnet.com/team/xml_wsspecs/xlang-c/default.htm, (2001).

15. Yang, J., Papazoglou, M.P.: Web Component: A Substrate for Web Service Reuse and Composition, CAiSE 2002, LNCS, Vol.2348, Springer-Verlag, 21–36(2002).

Author Index

Lecture Notes in Computer Science

For information about Vols. 1–3042

please contact your bookseller or Springer-Verlag